ACCADEMIA NAZIONALE DEI LINCEI
SCUOLA NORMALE SUPERIORE DI PISA

LEZIONI FERMIANE

★

CLASSE DI SCIENZE

SHMUEL AGMON

SPECTRAL PROPERTIES OF SCHRÖDINGER OPERATORS AND SCATTERING THEORY

PISA
MCMLXXV

Annali Scuola Normale Superiore - Pisa
Classe di Scienze
Serie IV - Vol. II, n. 2 (1975)

Spectral Properties of Schrödinger Operators and Scattering Theory (*) (**).

SHMUEL AGMON (***)

1. – Introduction.

We propose to discuss here certain spectral properties of Schrödinger operators $-\Delta + V(x)$ (Δ the Laplacian and V a potential) which have applications to scattering theory.

We denote by H_0 the self-adjoint realization of $-\Delta$ in $L^2(\boldsymbol{R}^n)$, with domain $\mathfrak{D}(H_0) = \mathcal{H}_2(\boldsymbol{R}^n)$ (¹). The complete spectral description of H_0 is obtained by application of the Fourier transform map $\mathcal{F}: L^2(\boldsymbol{R}^n) \to L^2(\boldsymbol{R}^n)$, defined by

$$(1.1)\qquad (\mathcal{F}u)(\xi) = \hat{u}(\xi) = (2\pi)^{-n/2}\int_{\boldsymbol{R}^n} u(x)\exp[-ix\cdot\xi]\,dx\,.$$

It follows that H_0 is unitarily equivalent to the operator of multiplication by $|\xi|^2$,

$$(1.2)\qquad (\mathcal{F}H_0)u(\xi) = |\xi|^2(\mathcal{F}u)(\xi)\,.$$

Consider a perturbation of H_0 given by the differential operator $-\Delta + V(x)$. We shall assume that $V(x)$ is a real function such that the multiplication operator $V: u \to V(x)u$ is H_0-compact (that is V is a compact operator from $\mathcal{H}_2(\boldsymbol{R}^n)$ into $L^2(\boldsymbol{R}^n)$). Under these conditions $-\Delta + V$ admits a unique self-adjoint realization in $L^2(\boldsymbol{R}^n)$ which we shall denote by H. It is well known that both operators H_0 and H possess the same domain of definition

(*) Fermi's Lectures supported by the Accademia Nazionale dei Lincei.

(**) Lectures given at the Scuola Normale Superiore of Pisa during March-April 1973 (revised and completed).

(***) The Hebrew University, Jerusalem.

Pervenuto alla Redazione il 30 Agosto 1974.

(¹) $\mathcal{H}_2$ denotes the Sobolev space of square integrable functions possessing square integrable derivatives up to the second order in the distribution sense.

and also the same essential spectrum which consists of the interval $[0, \infty)$.

We recall some notions from scattering theory. The wave operators of the pair (H, H_0) are defined as the strong limits

$$W_\pm = s - \lim_{t\to\pm\infty} \exp[itH]\exp[-itH_0]\,.$$

When the wave operators exist they yield isometries from $L^2(\boldsymbol{R}^n)$ into $L^2(\boldsymbol{R}^n)$. The wave operators are said to be complete if

$$\text{Range } W_+ = \text{Range } W_- = L^2(\boldsymbol{R}^n)_{\text{ac}}\,.$$

Here $L^2(\boldsymbol{R}^n)_{\text{ac}}$ denotes the absolute continuity subspace of $L^2(\boldsymbol{R}^n)$ with respect to H. (We recall that if A is a self-adjoint operator in a Hilbert space having a spectral resolution $\{E_\lambda\}$, then the absolute continuity subspace with respect to A consists of those vectors f for which $(E_\lambda f, f)$ is an absolutely continuous function of λ on $\boldsymbol{R}$, (see Kato [7], p. 516).)

When the wave operators exist and are complete one defines the scattering operator S by the relation: $S = W_+^* W_-$. S is a unitary operator on $L^2(\boldsymbol{R}^n)$ and one finds in a general situation, taking Fourier transform and introducing polar coordinates $\xi = (k, \omega)$, that

$$\mathcal{F}(Sf)(k, \omega) = \mathcal{S}(k)(\mathcal{F}f)(k, \cdot)(\omega)\,,$$

$f \in L^2(\boldsymbol{R}^n)$, where $\mathcal{S}(k)$ is an operator valued function defined for $k > 0$, taking its values in the class of unitary operators on $L^2(\Sigma)$, $\Sigma = \{\omega: |\omega| = 1, \omega \in \boldsymbol{R}^n\}$. One refers to $\mathcal{S}(k)$ as the *scattering matrix*.

The following problems arise.

PROBLEM I. *What are the conditions on V which ensure the existence and completeness of the wave operators.*

PROBLEM II. *What are the conditions which ensure the existence of the scattering matrix and what are the properties* (2) *of the operator valued function* $\mathcal{S}(k)$.

There is an extensive literature dealing with these problems. In particular the first problem was investigated by many authors who gave solutions which apply to different overlapping classes of potentials (see Kato [8; 9], Kato and Kuroda [10], Kuroda [11-14], Rejto [18], and other references given there).

(2) For example, it is of interest to give conditions which ensure that $\mathcal{S}(k)$ is an analytic operator valued function of k.

Among the various methods introduced in scattering theory there is an interesting method based on eigenfunction expansions. The method was introduced by Ikebe in [5] where he derived an eigenfunction expansion theorem, suitable for applications to scattering theory, for a class of Schrödinger operators in $\boldsymbol{R}^3$ (see also Povzner [17]). Ikebe's results were extended to $\boldsymbol{R}^n$ by Thoe [23], and were further improved by Alsholm and Schmidt [2] (3).

The main object of this study is to derive an eigenfunction expansion theorem for Schrödinger operators, which is useful as a tool in scattering theory, under minimal decay assumptions on the potential. Precisely we propose to establish the theorem under the condition that

$$\sup_{x\in R^n}\left[(1+|x|)^{2+2\varepsilon}\int\limits_{|y-x|\leqslant 1}|V(y)|^2\,|y-x|^{-n+\mu}\,dy\right]<\infty \tag{1.3}$$

for some $\varepsilon>0$ and μ satisfying $0<\mu<4$. Note that (1.3) holds in particular for V which at infinity verifies the condition:

$$V(x)=O\big(|x|^{-1-\varepsilon}\big)\qquad \text{as } |x|\to\infty, \tag{1.4}$$

and which locally verifies the condition: $V\in L^p_{\mathrm{loc}}(\boldsymbol{R}^n)$ with $p=2$ for $n\leqslant 3$, $p>n/2$ for $n\geqslant 4$. We observe here that the decay condition (1.4) is weaker than decay conditions imposed on V in the previously mentioned works on the eigenfunction expansion theorem, where it was (essentially) assumed that (4)

$$V(x)=O\big(|x|^{-(n+1)/2-\varepsilon}\big)\qquad \text{as } |x|\to\infty .$$

As an application of the eigenfunction expansion theorem we shall show that under condition (1.3) the wave operators exist and are complete (this result was recently proved by Greifenegger, Jörgen, Weidmann and Winkler [4]; see also Schechter [20]). We shall also show that the scattering matrix $S(k)$ exists as a continuous operator valued function and that, for a fixed k, $S(k)-I$ is a compact operator.

The plan of the paper is as follows. After introducing some notation, we begin in section 3 our study of Schrödinger operators. We consider an operator

(3) See also Shenk and Thoe [22], Schulenberger and Wilcox [21], where the eigenfunction expansion method is applied to other problems.

(4) This statement refers to simple pointwise decay conditions only. An integral decay condition on V which is given in [2] is not quite comparable with (1.3).

$H = -\Delta + V$ with potential V of class SR (a class which contains in particular all functions verifying condition (1.3)). We show that the positive point spectrum of H is a discrete set in $\boldsymbol{R}_+$. Eigenfunctions which correspond to positive eigenvalues are shown to decay rapidly as $|x| \to \infty$ (a well known result in the case of negative eigenvalues). This property is shown to hold also for certain generalized eigenfunctions, thus proving that such generalized eigenfunctions are necessarily proper eigenfunctions. In section 4 we establish the *limiting absorption principle*. That is, roughly speaking, we show that in some (optimal) topology the operator valued function $(H - z)^{-1}$, defined for non-real z, admits continuous boundary values on both sides of the positive axis (excluding the discrete set of eigenvalues). The limiting absorption principle is a basic tool in our study. Using this tool we introduce in section 5 the generalized eigenfunctions (distorted plane waves) for the class of Schrödinger operators with potentials verifying condition (1.3). Our main result, the eigenfunction expansion theorem, is proved in section 6. The applications of the eigenfunction expansion theorem to scattering theory are given in section 7.

There are three appendices where we prove some more special technical results. In particular in Appendix A we establish the a-priori weighted L^2 estimates which are used to prove our version of the limiting absorption principle (see Lemma 4.1). We go here beyond the needs of the present study and establish the estimates for the general class of partial differential operators with constant coefficients of *principal type*. In this connection we note that with the aid of these estimates one can easily extend the results of this paper (and in particular the eigenfunction expansion theorem) to a general class of self-adjoint operators H which are self-adjoint realizations of higher order elliptic operators. More precisely, all our results can be extended to the case where H is a self-adjoint realization in $L^2(\boldsymbol{R}^n)$ of a formally self-adjoint elliptic operator $P(x, D)$ of the form: $P(x, D) = P(D) + V(x, D)$, where $P(D)$ is an elliptic operator with constant coefficients of order m, and where $V(x, D) = \sum_{|\alpha| \leq m} V_\alpha(x) D^\alpha$ is a differential operator of order m with continuous top order coefficients satisfying condition (1.4), and with lower order coefficients V_α, $|\alpha| \leq m - 1$, satisfying condition (1.3) with $\mu = \mu_\alpha$, $0 < \mu_\alpha < 2(m - |\alpha|)$.

2. – Notation and definitions.

In our study we shall find it convenient to use various weighted L^2 spaces. For any real s we shall denote by $L^{2,s}(\boldsymbol{R}^n)$ the space of all complex valued functions on $\boldsymbol{R}^n$ defined by

$$L^{2,s}(\boldsymbol{R}^n) = \{u(x)\colon (1+|x|^2)^{s/2}u(x) \in L^2(\boldsymbol{R}^n)\}\,,$$

$x = (x_1, \ldots, x_n)$, $|x|^2 = x_1^2 + \ldots + x_n^2$ In $L^{2,s}$ we introduce the norm:

$$\|u\|_{0,s} = \|(1+|x|^2)^{s/2}u\|_{L^2(\boldsymbol{R}^n)}\,.$$

More generally, we shall consider weighted Sobolev L^2 spaces $\mathcal{H}_{m,s}(\boldsymbol{R}^n)$, defined for any integer $m \geqslant 0$ and real s by

$$\mathcal{H}_{m,s}(\boldsymbol{R}^n) = \{u(x)\colon D^\alpha u \in L^{2,s}(\boldsymbol{R}^n)\,, \qquad 0 \leqslant |\alpha| \leqslant m\}\,.$$

Here derivatives $D^\alpha u$ are taken in the distribution sense,

$$D^\alpha = D_1^{\alpha_1} \ldots D_n^{\alpha_n}\,, \qquad D_j = -\sqrt{-1}\,\frac{\partial}{\partial x_j}\,,$$

$\alpha = (\alpha_1, \ldots, \alpha_n)$ denoting a multi-index of order $|\alpha| = \alpha_1 + \ldots + \alpha_n$. $\mathcal{H}_{m,s}(\boldsymbol{R}^n)$ is a Hilbert space under the norm:

$$\|u\|_{m,s} = \Big(\sum_{|\alpha| \leqslant m} \|D^\alpha u\|_{0,s}^2\Big)^{\frac{1}{2}}\,.$$

The spaces $\mathcal{H}_{m,0}(\boldsymbol{R}^n)$, which are the usual L^2 Sobolev spaces of order m, will also be denoted by $\mathcal{H}_m(\boldsymbol{R}^n)$. As is well known one defines the Sobolev spaces $\mathcal{H}_m(\boldsymbol{R}^n)$ for any *real* m as follows. $\mathcal{H}_m(\boldsymbol{R}^n)$ is the completion of $C_0^\infty(\boldsymbol{R}^n)$ under the norm:

$$\|u\|_m = \Big(\int_{\boldsymbol{R}^n} |\hat{u}(\xi)|^2(1+|\xi|^2)^m\, d\xi\Big)^{\frac{1}{2}}\,.$$

We shall have the occasion to use the well known fact that a function in $\mathcal{H}_m(\boldsymbol{R}^n)$, for $m > \frac{1}{2}$, has a trace on any sufficiently smooth $n-1$ dimensional manifold imbedded in $\boldsymbol{R}^n$. More precisely, we shall use the following special

TRACE THEOREM ([5]). *Let Γ be a C^∞ compact $n-1$ dimensional manifold imbedded in $\boldsymbol{R}^n$. Let $d\sigma$ be the measure induced on Γ by the Lebesgue measure dx, and denote by $L^2(\Gamma)$ the class of L^2 functions on Γ with respect to the measure $d\sigma$.*

([5]) E.g. Lions-Magenes [15], p. 44.

For any given $m > \frac{1}{2}$, *there exists a bounded linear map*

$$\tau: \mathcal{H}_m(\boldsymbol{R}^n) \to L^2(\Gamma)$$

such that

$$\tau u = u|_\Gamma \quad \text{for } u \in \mathcal{H}_m(\boldsymbol{R}^n) \cap C(\boldsymbol{R}^n) .$$

One refers to τu as the trace of u on Γ. In particular, if $\tau u = 0$ one says that $u = 0$ on Γ in the trace sense.

Let X, Y be two Banach spaces. We shall denote by $B(X, Y)$ the space of all bounded linear operators from X into Y. As usual we shall consider $B(X, Y)$ as a Banach space whose norm is given by the operator norm.

3. – Schrödinger operators. The positive point spectrum and a decay property of eigenfunctions.

In this and in the following section we shall consider Schrödinger operators with potentials of class SR. This general class of potentials is defined as follows.

DEFINITION 3.1. *A real function* $V(x) \in L^2_{\text{loc}}(\boldsymbol{R}^n)$ *is said to belong to the class* SR *(short range) if, for some* $\varepsilon > 0$, *the multiplication mapping*:

$$u(x) \to (1 + |x|)^{1+\varepsilon}\, V(x)\, u(x)$$

defines a compact operator from $\mathcal{H}_2(\boldsymbol{R}^n)$ *into* $L^2(\boldsymbol{R}^n)$.

REMARK 1. If V is of class SR, then for some $\varepsilon > 0$ and any real s the multiplication operator:

$$M_V: u(x) \to V(x)\, u(x)$$

is a compact operator from $\mathcal{H}_{2,s}(\boldsymbol{R}^n)$ into $L^{2,s+1+\varepsilon}(\boldsymbol{R}^n)$. Indeed, this is obvious for $s = 0$. For a general s the result follows from the special case noting that the mapping: $u \to (1 + |x|^2)^{-s/2} u$ defines a bounded operator from $\mathcal{H}_{m,r}(\boldsymbol{R}^n)$ into $\mathcal{H}_{m,r+s}(\boldsymbol{R}^n)$ for any real r, s and $m = 0, 1, 2, \ldots$.

REMARK 2. If V is a real function verifying condition (1.3) then V is of class SR. This follows from a well known result (see Schechter [19], Ch. 6) by which condition (1.3) implies that the map: $u \to (1 + |x|)^{1+\varepsilon'}\, Vu$, defines a compact operator from $\mathcal{H}_2(\boldsymbol{R}^n)$ into $L^2(\boldsymbol{R}^n)$ for any $0 < \varepsilon' < \varepsilon$.

Consider a Schrödinger differential operator $-\Delta + V(x)$ with potential $V(x)$ of class SR. As before denote by H_0 the self-adjoint realization of $-\Delta$ in $L^2(\boldsymbol{R}^n)$, $\mathfrak{D}(H_0) = \mathcal{H}_2(\boldsymbol{R}^n)$. Denote by H the operator: $H_0 + V$ in $L^2(\boldsymbol{R}^n)$ with domain $\mathfrak{D}(H) = \mathcal{H}_2(\boldsymbol{R}^n)$ (V the multiplication operator by $V(x)$). Since V is a symmetric operator in $L^2(\boldsymbol{R}^n)$ which by our assumption is H_0-compact, it follows from a well known theorem (e.g. [7]; p. 287) that H is a self-adjoint operator in $L^2(\boldsymbol{R}^n)$. It could also be shown that H is the *unique* self-adjoint realization in $L^2(\boldsymbol{R}^n)$ of the differential operator $-\Delta + V(x)$ and that H is semi-bounded. Let $\sigma(H)$ be the spectrum of H. It is well known that $\sigma(H) = [0, \infty) \cup \{\lambda_j\}$ where $[0, \infty)$ is the essential spectrum of H and $\{\lambda_j\}$ is a discrete set of negative eigenvalues with a finite multiplicity, having zero as its only limit point.

We note that the results just described do not require the full strength of our assumption on V. For instance, it is well known that all the results hold under the weaker assumption that V is H_0-compact. However, the assumption that V is of class SR will be used in an essential way in the following results.

With some abuse of notation we shall sometimes use the notation $-\Delta + V$ to denote the operator H.

We denote by $e_+(H)$ the set of all positive eigenvalues of H.

THEOREM 3.1. *$e_+(H)$ is a discrete set on the real line. The only possible limit points of $e_+(H)$ on the extended line are the points $\lambda = 0$ and $\lambda = +\infty$. Every point in $e_+(H)$ is an eigenvalue of a finite multiplicity.*

REMARK. Under more stringent conditions on V than those assumed here it could be shown that the set $e_+(H)$ is empty (e.g. [2; Appendix 3]). It is an open question whether $e_+(H)$ is empty for any potential V of class SR(6).

For the proof of Theorem 3.1 we need the following

THEOREM 3.2. *Let $f(x) \in \mathcal{H}_s(\boldsymbol{R}^n)$ for some $s > \frac{1}{2}$. Suppose that $f(x) = 0$ on a sphere $|x| = k$ in the trace sense, and let $K^{-1} \leqslant k \leqslant K$, K some positive constant. For any multi-index α with $0 \leqslant |\alpha| \leqslant 2$, set*

$$v_\alpha(x) = \frac{x^\alpha f(x)}{|x^2| - k^2}, \qquad x^\alpha = x_1^{\alpha_1} \dots x_n^{\alpha_n}.$$

(6) The proof that $e_+(H)$ is empty requires, among other things, the validity of the unique continuation property for solutions of the equation $-\Delta u + Vu = \lambda u$. This may indicate that the property: $e_+(H) = \emptyset$ need not hold for V which is too singular.

Then $v_\alpha \in \mathcal{H}_{s-1}(\boldsymbol{R}^n) \cap L^1_{\text{loc}}(\boldsymbol{R}^n)$, *and*

$$\|v_\alpha\|_{s-1} \leqslant C\|f\|_s \tag{3.1}$$

where C is a constant depending only on s and K.

The proof of Theorem 3.2 (in a more general set up) is given in Appendix B. We turn to the

PROOF OF THEOREM 3.1. We shall show that if $u \in \mathcal{H}_2(\boldsymbol{R}^n)$ is an eigenfunction of H corresponding to a positive eigenvalue λ, $0 < a \leqslant \lambda \leqslant b$, then $u \in \mathcal{H}_{2,\varepsilon}(\boldsymbol{R}^n)$ for some $\varepsilon > 0$ depending only on V, and that

$$\|u\|_{2,\varepsilon} \leqslant C\|u\|_0 \tag{3.2}$$

where C is a constant depending only on V, a and b. The theorem is an easy corollary of the estimate (3.2). Indeed, by a variant of Rellich's compactness theorem (see [1], p. 30) it follows that the injection map of $\mathcal{H}_{2,\varepsilon}(\boldsymbol{R}^n)$ in $L^2(\boldsymbol{R}^n)$ is compact for any $\varepsilon > 0$. Hence, any orthonormal set of functions $\{u_j\}$ in $L^2(\boldsymbol{R}^n)$, which is also a bounded set in $\mathcal{H}_{2,\varepsilon}(\boldsymbol{R}^n)$, is necessarily a finite set. This and (3.2) clearly imply that H has only a finite number of eigenvalues in $[a, b]$ and that the multiplicity of each eigenvalue is finite.

To prove (3.2), observe first that the estimate (3.2) holds for $\varepsilon = 0$. Indeed, let $R(i) = (H - i)^{-1}$. The relation $Hu = \lambda u$ implies that $u = (\lambda - i)R(i)u$. Since $R(i)$ is also a bounded operator from $L^2(R^n)$ into $\mathfrak{D}(H) = \mathcal{H}_2(R^n)$, it follows that

$$\|u\|_2 = \|(\lambda - i)R(i)u\|_2 \leqslant C'\|u\|_0 \tag{3.2$'$}$$

where C' is a constant depending only on V and b.

We set $g(x) = -V(x)u(x)$. Using our assumption that V is of class SR, it follows from Remark 1 (following Definition 3.1) that the multiplication operator V is a compact operator from $\mathcal{H}_{2,s}(\boldsymbol{R}^n)$ into $L^{2,s+1+\varepsilon}(\boldsymbol{R}^n)$, for any real s and a certain fixed $\varepsilon > 0$. Hence, in particular, we have

$$\|Vw\|_{0,s+1+\varepsilon} \leqslant \gamma\|w\|_{2,s} \quad \text{for } \forall w \in \mathcal{H}_{2,s}(\boldsymbol{R}^n) \tag{3.3}$$

and any s, where γ is a constant depending only on V and s. Applying (3.3) with $w = u$ and $s = 0$, using (3.2)$'$, we conclude that $g \in L^{2,1+\varepsilon}(\boldsymbol{R}^n)$ and that

$$\|g\|_{0,1+\varepsilon} \leqslant \gamma\|u\|_2 \leqslant C_1\|u\|_0, \tag{3.4}$$

C_j denote constants which depend only on V, ε, a and b.

Now, since $g(x) \in L^{2,1+\varepsilon}(\boldsymbol{R}^n)$, it follows by Fourier transform that $\hat{g}(\xi) \in \mathcal{H}_{1+\varepsilon}(\boldsymbol{R}^n)$. Since $u(x)$ satisfies the differential equation: $(\Delta + \lambda)u = Vu$, we also have, by Fourier transform,

$$(3.5) \qquad (|\xi|^2 - \lambda)\,\hat{u}(\xi) = -\mathcal{F}(Vu)(\xi) = \hat{g}(\xi)\,.$$

From (3.5) it follows that

$$(3.6) \qquad \hat{g}(\xi)\big|_{|\xi|=\sqrt{\lambda}} = 0 \quad \text{in the trace sense}\,.$$

To see this (7) we use the formula:

$$(3.7) \qquad \lim_{h\to+0} \frac{1}{2h} \int\limits_{||\xi|-\sqrt{\lambda}|\leqslant h} \hat{g}(\xi)\,\varphi(\xi)\,d\xi = \int\limits_{|\xi|=\sqrt{\lambda}} (\tau\hat{g})(\xi)\cdot\varphi(\xi)\,d\sigma$$

which holds if $\hat{g} \in \mathcal{H}_s(\boldsymbol{R}^n)$ with $s > \frac{1}{2}$ and $\varphi \in C(\boldsymbol{R}^n)$. Here $\tau\hat{g}$ denotes the trace of $\hat{g}$ on the sphere $|\xi| = \sqrt{\lambda}$, and $d\sigma$ is the induced measure on the sphere. (Indeed, (3.7) is immediate if $\hat{g}$ is also continuous. In the general case the result follows from this by a density argument, using the continuity of the trace map τ.) Combining (3.5) and (3.7), it follows readily that

$$\int\limits_{|\xi|=\sqrt{\lambda}} (\tau\hat{g})(\xi)\cdot\varphi(\xi)\,d\sigma = 0 \qquad \text{for } \forall\varphi \in C(\boldsymbol{R}^n)\,,$$

which implies (3.6).

Finally, apply Theorem 3.2 to the function $f(x) = \hat{g}(x)$. Since $\hat{g} \in \mathcal{H}_{1+\varepsilon}(\boldsymbol{R}^n)$, and since $\hat{u}(\xi) = \hat{g}(\xi)(|\xi|^2 - \lambda)^{-1}$, it follows from Theorem 3.2 that $\xi^\alpha\hat{u}(\xi) \in \mathcal{H}_\varepsilon(\boldsymbol{R}^n)$ for $|\alpha| \leqslant 2$. This in turn implies that $u \in \mathcal{H}_{2,\varepsilon}(\boldsymbol{R}^n)$. Moreover, combining (3.1) (with $s = 1 + \varepsilon$) and (3.4), we find that

$$\|u\|_{2,\varepsilon} = \Big(\sum_{|\alpha|\leqslant 2} \|\xi^\alpha\hat{u}\|_\varepsilon^2\Big)^{\frac{1}{2}} \leqslant C_2\|\hat{g}\|_{1+\varepsilon}$$
$$= C_2\|g\|_{0,1+\varepsilon} \leqslant C_3\|u\|_0\,,$$

which yields (3.2). The proof of the theorem is now complete.

In proving Theorem 3.1 we have obtained as a by-product the result that any eigenfunction of H, which corresponds to a positive eigenvalue λ, belongs to the weighted Sobolev class $\mathcal{H}_{2,\varepsilon}(\boldsymbol{R}^n)$ for some $\varepsilon > 0$. As a matter

(7) (3.6) does not follow from (3.5) immediately since we know only that $\hat{u} \in L^2(\boldsymbol{R}^n)$.

of fact one can show more, namely that $u \in \mathcal{H}_{2,s}(\boldsymbol{R}^n)$ for any $s > 0$. This useful decay property could also be shown to hold for certain improper eigenfunctions of H, which to start with are not assumed to be in the domain of H. The general result is the following

THEOREM 3.3. *Let $V(x)$ be a real function on $\boldsymbol{R}^n$ such that $(1 + |x|)^\delta V(x)$ belongs to the class SR for some $\delta > 0$. Let $u(x) \in \mathcal{H}_2^{\text{loc}}(\boldsymbol{R}^n)$ be a solution of differential equation*

$$-\Delta u + V(x)u = \lambda u \qquad \text{in } \boldsymbol{R}^n \text{ (}^8\text{)},$$

λ a positive number. Suppose that $u \in \mathcal{H}_{2,s_0}(\boldsymbol{R}^n)$ for some $s_0 > -\frac{1}{2} - \delta$. Consider u as a tempered distribution acting on $\mathcal{S}(\boldsymbol{R}^n)$ (⁹), and let $\hat{u}$ be the distributional Fourier transform of u $(\hat{u} \in \mathcal{S}'(\boldsymbol{R}^n))$. If $\hat{u} \in L^1_{\text{loc}}(\boldsymbol{R}^n)$; then $u \in \mathcal{H}_{2,s}(\boldsymbol{R}^n)$ for any real s.

PROOF. To prove the theorem it suffices to show that under the conditions of the theorem it follows that $u \in \mathcal{H}_{2,s_0+\delta}(\boldsymbol{R}^n)$. For then it would follow that the conditions of the theorem hold with s_0 replaced by $s_0 + \delta$. Applying the same result it would then follow that $u \in \mathcal{H}_{2,s_0+2\delta}(\boldsymbol{R}^n)$. Continuing in this manner we shall find that $u \in \mathcal{H}_{2,s_0+j\delta}(\boldsymbol{R}^n)$ for $j = 1, 2, \ldots$, which is the desired result.

Now, to prove that $u \in \mathcal{H}_{2,s_0+\delta}(\boldsymbol{R}^n)$ we repeat the same argument used in the proof of (3.2). We set $g(x) = -V(x)u(x)$. It follows from our assumptions, using (3.3) with $w = u$ and $\varepsilon = \delta$, that $g(x) \in L^{2,s_0+1+\delta}(\boldsymbol{R}^n)$. By Fourier transform we have that $\hat{g}(\xi) \in \mathcal{H}_{s_0+1+\delta}(\boldsymbol{R}^n)$ and that the relation (3.5) holds. Since $\hat{u} \in L^1_{\text{loc}}(\boldsymbol{R}^n)$ it follows from (3.5), with the aid of (3.7) as before, that the trace relation (3.6) holds (here we use our assumption that $s_0 + 1 + \delta > \frac{1}{2}$). Apply now Theorem 3.2 to the function $f(x) = \hat{g}(x)$. Since $\hat{u}(\xi) = \hat{g}(\xi)(|\xi|^2 - \lambda)^{-1}$, and since $\hat{g} \in \mathcal{H}_{s_0+1+\delta}\boldsymbol{R}^n)$, it follows from the theorem that $\xi^\alpha \hat{u}(\xi) \in \mathcal{H}_{s_0+\delta}(\boldsymbol{R}^n)$ for $|\alpha| \leq 2$. This implies that $u \in \mathcal{H}_{2,s_0+\delta}(\boldsymbol{R}^n)$, thus completing the proof of the theorem.

4. – The limiting absorption principle.

Let $R(z) = (H - z)^{-1}$ be the resolvent of H defined for $z \notin \sigma(H)$. It is of interest to know whether $R(z)$ assumes in some sense boundary values on the positive axis $\boldsymbol{R}_+ = \{\lambda: \lambda \in \boldsymbol{R}, \lambda > 0\}$, obtained as limits of $R(z)$ as

(⁸) Here and elsewhere solutions of the Schrödinger differential equation should be taken in the distribution sense.

(⁹) $\mathcal{S}(\boldsymbol{R}^n)$ denotes the Schwartz class of rapidly decreasing C^∞ functions on $\boldsymbol{R}^n$.

$z \to \lambda$ through points in one of the half-planes: $\boldsymbol{C}_{\pm} = \{z : z \in \boldsymbol{C},\ \pm \operatorname{Im} z > 0\}$. Since $\boldsymbol{R}_+ \subset \sigma(H)$, it is clear that such limits cannot exist in the uniform operator topology of $B(L^2, L^2)$ (or in any weaker L^2 topology). However, as we shall see, such limits do exist if one considers $R(z)$ as an operator valued function with values in $B(L^{2,s}, L^{2,-s})$ (or even in $B(L^{2,s}, \mathcal{H}_{2,-s})$) for any $s > \frac{1}{2}$. This result, which following an accepted terminology we call the *limiting absorption principle*, is the main result of this section. It has a basic role in our subsequent study of the eigenfunction expansion theorem.

The proof of the limiting absorption principle is long and we shall arrive at the final result through intermediate steps. The first step in the proof consists in verifying the result for the special case of the unperturbed operator $H_0 = -\Delta$.

THEOREM 4.1. *Let $R_0(z) = (H_0 - z)^{-1}$. Consider $R_0(z)$ as an analytic operator valued function on $\boldsymbol{C} \setminus \overline{\boldsymbol{R}}_+$ with values in $B(L^{2,s}, \mathcal{H}_{2,-s})$, $s > \frac{1}{2}$. Then*

(i) *For any $\lambda \in \boldsymbol{R}_+$, the following two limits exist in the uniform operator topology of $B(L^{2,s}, \mathcal{H}_{2,-s})$:*

$$\lim_{\substack{z \to \lambda \\ \pm Im z > 0}} R_0(z) = R_0^{\pm}(\lambda)\,. \tag{4.1}$$

(ii) *For any $f \in L^{2,s}(\boldsymbol{R}^n)$ and $\lambda \in \boldsymbol{R}_+$ the function $u = R_0^{\pm}(\lambda) f$ verifies the differential equation*

$$(-\Delta - \lambda) u = f\,. \tag{4.2}$$

The following identity holds [10]:

$$\operatorname{Im} \langle R_0^{\pm}(\lambda) f, f \rangle = \pm \frac{\pi}{2\sqrt{\lambda}} \int_{|\xi| = \sqrt{\lambda}} |(\tau \hat{f})(\xi)|^2 \, d\sigma \tag{4.3}$$

where $\tau \hat{f}$ denotes the trace of $\hat{f} = \mathcal{F} f$ on the sphere $|\xi| = \sqrt{\lambda}$.

The crux of Theorem 4.1 are certain weighted L^2 estimates for the operator $\Delta + z$ which we state as

LEMMA 4.1. *Let $s > \frac{1}{2}$, $K > 1$. There exists a constant C depending only*

[10] For any pair of functions f, g on $\boldsymbol{R}^n$ with $f\bar{g} \in L^1(\boldsymbol{R}^n)$, we define

$$\langle f, g \rangle = \int_{\boldsymbol{R}^n} f(x) \overline{g(x)} \, dx\,.$$

on s, n and K, such that

$$\|u\|_{2,-s} \leqslant C\|(\Delta + z)u\|_{0,s} \tag{4.4}$$

for $\forall u \in \mathcal{H}_2(\boldsymbol{R}^n)$ *and* $\forall z \in \boldsymbol{C}$ *such that* $K^{-1} \leqslant |z| \leqslant K$.

Lemma 4.1 is proved in Appendix A (it follows as a special case of estimates established for general elliptic operators, see Theorem A1). With the aid of the lemma we give now the

PROOF OF THEOREM 4.1. Let $f, g \in L^{2,s}(\boldsymbol{R}^n)$, $s > \frac{1}{2}$. We shall first show that the function $F(z) = \langle R_0(z)f, g\rangle$ which is an analytic function of z in $\boldsymbol{C}\diagdown\overline{\boldsymbol{R}}_+$, has continuous boundary values on both edges of $\boldsymbol{R}_+$. Now, by Lemma 4.1, we have

$$\|R_0(z)f\|_{2,-s} \leqslant C\|f\|_{0,s} \tag{4.5}$$

for $\forall f \in L^{2,s}(\boldsymbol{R}^n)$ and $\forall z \in \boldsymbol{C}\diagdown\boldsymbol{R}_+$ such that $K^{-1} \leqslant |z| \leqslant K$, (any fixed $K > 1$). This implies that with the same constant C, we have

$$|\langle R_0(z)f, g\rangle| \leqslant C\|f\|_{0,s}\|g\|_{0,s}\,, \tag{4.5$'$}$$

for $\forall f$, $g \in L^{2,s}(\boldsymbol{R}^n)$ and $\forall z \in \boldsymbol{C}\diagdown\boldsymbol{R}_+$ such that $K^{-1} \leqslant |z| \leqslant K$. The uniform bound (4.5)$'$ implies that in order to prove that $\langle R_0(z)f, g\rangle$ assumes continuous boundary values on both sides of $\boldsymbol{R}_+$ for any f, $g \in L^{2,s}(\boldsymbol{R}^n)$, it suffices to show that this is true for f and g in some dense set in $L^{2,s}(\boldsymbol{R}^n)$.

Let f, $g \in C_0^\infty(\boldsymbol{R}^n)$. By Fourier transform (using Parseval's formula and changing to polar coordinates) it follows that

$$\langle R_0(z)f, g\rangle = \int_{\boldsymbol{R}^n} \frac{\hat{f}(\xi)\overline{\hat{g}(\xi)}}{|\xi|^2 - z}\,d\xi = \frac{1}{2}\int_0^\infty \frac{t^{(n-2)/2}}{t - z}\left(\int_{|\omega|=1} \hat{f}(\sqrt{t}\omega)\,\overline{\hat{g}(\sqrt{t}\omega)}\,d\omega\right)dt \tag{4.6}$$

By well known continuity properties of Cauchy type integrals, it follows from (4.6) that the function $\langle R_0(z)f, g\rangle$ has continuous boundary values on both edges of $\boldsymbol{R}_+$ given by

$$\lim_{\substack{z\to\lambda\in\boldsymbol{R}_+\\ \pm\operatorname{Im} z>0}} \langle R_0(z)f, g\rangle = \pm\frac{\pi i}{2\sqrt{\lambda}} \int_{|\xi|=\sqrt{\lambda}} \hat{f}(\xi)\overline{\hat{g}(\xi)}\,d\sigma + \text{p.v.}\int_{\boldsymbol{R}^n} \frac{\hat{f}(\xi)\overline{\hat{g}(\xi)}}{|\xi|^2 - \lambda}\,d\xi\,, \tag{4.7}$$

the last singular integral being taken in the principal value sense.

From (4.7) and our previous remarks it follows that $\langle R_0(z)f, g\rangle$ admits continuous boundary values on both sides of $\boldsymbol{R}_+$ for any f, $g \in L^{2,s}(\boldsymbol{R}^n)$.

This in turn means that for $f \in L^{2,s}(\boldsymbol{R}^n)$:

$$\exists w - \lim_{\substack{z \to \lambda \in \boldsymbol{R}^+ \\ \pm \operatorname{Im} z > 0}} R_0(z) f \qquad \text{in } L^{2,-s}(\boldsymbol{R}^n) . \tag{4.8}$$

Moreover, since (by (4.5)) $R_0(z)f$ is also bounded near λ when considered as a function with values in $\mathcal{H}_{2,-s}(\boldsymbol{R}^n)$, it follows (using the weak compactness of the unit ball in $\mathcal{H}_{2,-s}(\boldsymbol{R}^n)$) that the limit (4.8) also exists as a weak limit in $\mathcal{H}_{2,-s}(\boldsymbol{R}^n)$. We shall define for any $f \in L^{2,s}(\boldsymbol{R}^n)$ and $\lambda \in \boldsymbol{R}_+$:

$$R_0^{\pm}(\lambda) f = w - \lim_{\substack{z \to \lambda \\ \pm \operatorname{Im} z > 0}} R_0(z) f \qquad \text{in } \mathcal{H}_{2,-s}(\boldsymbol{R}^n) . \tag{4.9}$$

It is clear that (4.9) defines an operator $R_0^{\pm}(\lambda) \in B(L^{2,s}, \mathcal{H}_{2,-s})$ for any $s > \frac{1}{2}$.

We are going to show that the weak boundary values $R_0^{\pm}(\lambda)$ of $R_0(z)$ on $\boldsymbol{R}_+$ just defined are actually assumed in the stronger sense (4.1). Before doing this let us observe that the relations described in part (ii) of the theorem are simple consequences of (4.9) and (4.7). Indeed, that $u = R_0^{\pm}(\lambda) f$ satisfies equation (4.2) (in the distribution sense) follows immediately from (4.9) and the relation: $\langle R_0(z)f, (-\Delta - \bar{z})\varphi \rangle = \langle f, \varphi \rangle$, which holds for $\forall \varphi \in C_0^\infty(\boldsymbol{R}^n)$ and $z \notin \boldsymbol{R}$. Formula (4.3) for $f \in C_0^\infty(\boldsymbol{R}^n)$ follows immediately from (4.7) (taking $f = g$). That (4.3) holds for all $f \in L^{2,s}(\boldsymbol{R}^n)$ follows from this by continuity, noting that both sides of (4.3) represent continuous quadratic functionals on $L^{2,s}(\boldsymbol{R}^n)$ for any $s > \frac{1}{2}$.

We continue with the proof of (i). Define the operator valued function $R_0^+(z)$ on $\tilde{C}_+ = \boldsymbol{C}_+ \cup \boldsymbol{R}_+$, and the operator valued function $R_0^-(z)$ on $\tilde{C}_- = \boldsymbol{C}_- \cup \boldsymbol{R}_+$ ($R_0^+(z)$ and $R_0^-(z)$ with values in $B(L^{2,s}, \mathcal{H}_{2,-s})$) as follows:

$$R_0^{\pm}(z) = R_0(z) \qquad \text{if } \pm \operatorname{Im} z > 0 ,$$

$$R_0^{\pm}(z) \qquad \text{is the operator defined by (4.9) if } z = \lambda \in \boldsymbol{R}_+ .$$

From the preceding it is clear that $R_0^{\pm}(z)$ is a weakly continuous operator valued function on $\tilde{C}_{\pm}$. We shall show that $R_0^{\pm}(z)$ is actually continuous on $\tilde{C}_{\pm}$ in the uniform operator topology of $B(L^{2,s}, \mathcal{H}_{2,-s})$. To this end observe that for u in $\mathcal{H}_{2,-s}(\boldsymbol{R}^n)$ the norm $\|u\|_{2,-s}$ is equivalent to the norm $\|(1+|x|^2)^{-s/2}u\|_2$. As is well known this last norm is equivalent to $\|(I-\Delta)(1+|x|^2)^{-s/2}u\|_0$ which in turn is equivalent to the norm $|||u||| = \|u\|_{0,-s} + \|\Delta u\|_{0,-s}$. From this it follows that in order to show that $R_0^{\pm}(z)$ is continuous on $\tilde{C}_{\pm}$ in the uniform operator topology of $B(L^{2,s}, \mathcal{H}_{2,-s})$, it suffices to show that both $R_0^{\pm}(z)$ and $\Delta R_0^{\pm}(z)$ are continuous in the uniform operator topology of $B(L^{2,s}, L^{2,-s})$. Since $-\Delta R_0^{\pm}(z) = I + z R_0^{\pm}(z)$, this reduces simply to showing that $R_0^{\pm}(z)$ is continuous on $\tilde{C}_{\pm}$ in the uniform operator topology of $B(L^{2,s}, L^{2,-s})$. In order to establish this result let us first observe that $R_0^{\pm}(z)$ is continuous on $\tilde{C}_{\pm}$ in the strong topology of $B(L^{2,s}, L^{2,-s})$, i.e. for any $z_0 \in \tilde{C}_{\pm}$ and $f \in L^{2,s}(\boldsymbol{R}^n)$

$$\exists s - \lim_{\substack{z \to z_0 \\ z \in \tilde{\boldsymbol{C}}_{\pm}}} R_0^{\pm}(z) f = R_0^{\pm}(z_0) f \qquad \text{in } L^{2,-s}(\boldsymbol{R}^n) . \tag{4.10}$$

Indeed, (4.10) is obvious if $z_0 \notin \boldsymbol{R}_+$. If $z_0 = \lambda \in \boldsymbol{R}_+$ then (4.10) follows from (4.9), using the compactness of the injection map of $\mathcal{H}_{2,-s'}(\boldsymbol{R}^n)$ in $L^{2,-s}(\boldsymbol{R}^n)$ for any $s' < s$ [11] (we also use the observation that $R_0(z) f \in \mathcal{H}_{2,-s'}$, and that the limit relation (4.9) holds in $\mathcal{H}_{2,-s'}$, for any $s' > \frac{1}{2}$).

Next we observe that if $\{z_j\} \subset \boldsymbol{C}_{\pm}$, and $\{f_j\} \subset L^{2,s}(\boldsymbol{R}^n)$, are sequences such that

$$\lim z_n = z_0 \in \boldsymbol{C}_{\pm} \qquad \text{and} \qquad w - \lim f_j = f \quad \text{in } L^{2,s}(\boldsymbol{R}^n) ,$$

then

$$s - \lim_{j \to \infty} R_0^{\pm}(z_j) f_j = R_0^{\pm}(z_0) f \qquad \text{in } L^{2,-s}(\boldsymbol{R}^n) . \tag{4.11}$$

Indeed, for any $g \in L^{2,s}(\boldsymbol{R}^n)$, we have

$$\lim_{j \to \infty} \langle R_0^{\mp}(z_j) f_j, g \rangle = \lim_{j \to \infty} \langle f_j, R_0^{\pm}(\bar{z}_j) g \rangle = \langle f, R_0^{\pm}(\bar{z}_0) g \rangle = \langle R_0^{\mp}(z_0) f, g \rangle \ (^{12}) ,$$

which shows that

$$w - \lim_{j \to \infty} R_0^{\pm}(z_j) f_j = R_0^{\pm}(z_0) f \qquad \text{in } L^{2,-s}(\boldsymbol{R}^n) . \tag{4.11$'$}$$

Since (by (4.5)) the sequence $\{R_0^{\pm}(z_j) f_j\}$ is bounded in $\mathcal{H}_{2,-s'}(\boldsymbol{R}^n)$ for any $s' > \frac{1}{2}$, and since the injection map of $\mathcal{H}_{2,-s'}(R^n)$ in $L_{2,-s}(R^n)$ is compact, for any $s' < s$, it follows from the existence of the weak limit $(4.11)'$ that (4.11) holds.

Finally, it follows from (4.11) that $R_0^{\pm}(z)$ is continuous on $\tilde{C}_{\pm}$ in the uniform operator topology of $B(L^{2,s}, L^{2,-s})$. Indeed, suppose by contradiction that this is not true. This would imply that there exist a sequence

[11] This follows from Rellich's compactness theorem.

[12] The relation: $\langle R_0^{\pm}(z) f, g \rangle = \langle f, R_0^{\mp}(\bar{z}) g \rangle$, $f, g \in L^{2,s}$, which is obvious when $z \notin \boldsymbol{R}_+$, is also valid by continuity when $z \in \boldsymbol{R}_+$.

$\{z_j\}\subset\tilde{C}_\pm$ with $z_j\to z_0\in\tilde{C}_\pm$, and a sequence $\{f_j\}\subset L^{2,s}(\boldsymbol{R}^n)$ with $\|f_j\|_{0,s}=1$, such that

$$\lim_{j\to\infty}\|(R_0^\pm(z_j)-R_0^\pm(z_0))f_j\|_{0,-s}>0\ . \tag{4.12}$$

Extracting if necessary a subsequence we may also assume that $\exists w-\lim f_j=f$ in $L^{2,s}(\boldsymbol{R}^n)$. Applying (4.11), it follows that

$$s-\lim R_0^\pm(z_j)f_j=R_0^\pm(z_0)f=s-\lim R_0^\pm(z_0)f_j\quad\text{in } L^{2,-s}(\boldsymbol{R}^n)\ .$$

This gives a contradiction (to (4.12)), proving the continuity of $R_0^\pm(z)$ on $\tilde{C}_\pm$ in the uniform operator topology of $B(L^{2,s},L^{2,-s})$. As was observed before this last result implies in its turn that $R_0^\pm(z)$ is continuous on $\tilde{C}_\pm$ in the uniform topology of $B(L^{2,s},\mathcal{H}_{2,-s})$. This yields (4.1) and completes the proof of the theorem.

We introduce the following

Definition 4.1. *A function $u\in\mathcal{H}_2^{\text{loc}}(\boldsymbol{R}^n)$ will be called a k-outgoing function (resp. k-incoming function) if for $k>0$ the following relation holds:*

$$u=R_0^+(k^2)f\qquad(\textit{resp. } u=R_0^-(k^2)f) \tag{4.13}$$

for some $f\in L^{2,s}(\boldsymbol{R}^n)$, $s>\frac12$. (Here $R_0^\pm(k^2)\in B(L^{2,s},\mathcal{H}_{2,-s})$ is defined by Theorem 4.1.)

Note that if u is an outgoing or an incoming function then $u\in\bigcap_{s>\frac12}\mathcal{H}_{2,-s}(\boldsymbol{R}^n)$. The following result will be needed later on.

Lemma 4.2. *Let $u\in\mathcal{H}_2^{\text{loc}}(\boldsymbol{R}^n)$ be a k-outgoing (k-incoming) function satisfying a differential equation of the form:*

$$-\Delta u+V(x)u=k^2u\quad\text{in } \boldsymbol{R}^n\ , \tag{4.14}$$

where V is a real function of class SR. Then $u\in\mathcal{H}_{2,s}(\boldsymbol{R}^n)$ for all real s.

Proof. We shall prove the lemma for u outgoing, the proof for u incoming is similar. By assumption $u=R_0^+(k^2)f$ for some $f\in L^{2,s_0}(\boldsymbol{R}^n)$, $s_0>\frac12$. This implies that $f=(-\Delta-k^2)u$ which, when compared with (4.14), gives that $f=-Vu$. Applying formula (4.3), using the last relation and the reality of V, we get

$$\int_{|\xi|=k}|(\tau\hat{f})(\xi)|^2d\sigma=\frac{2k}{\pi}\operatorname{Im}\langle R_0^+(k^2)f,f\rangle=-\frac{2k}{\pi}\operatorname{Im}\langle u,Vu\rangle=0\ . \tag{4.15}$$

From (4.15) it follows that $\hat{f}(\xi)=0$ on the sphere $|\xi|=k$ (trace sense; note that $\hat{f}\in\mathcal{H}_{s_0}$). Hence, applying Theorem 3.2 to $\hat{f}$ it follows that

$$\hat{f}(\xi)(|\xi|^2-k^2)^{-1}\in L^1_{\text{loc}}(\boldsymbol{R}^n)\,. \tag{4.16}$$

Next we show that

$$\hat{u}=\hat{f}(\xi)(|\xi|^2-k^2)^{-1} \tag{4.16$'$}$$

where $\hat{u}\in\mathcal{S}'(\boldsymbol{R}^n)$ is the distributional Fourier transform of u (note that u is a tempered distribution since $u\in L^{2,-s_0}(\boldsymbol{R}^n)$). Indeed, let $g\in\mathcal{S}(\boldsymbol{R}^n)$. Since $u=\lim\limits_{\varepsilon\to+0}R_0(k^2+i\varepsilon)f$ in $L^{2,-s_0}(\boldsymbol{R}^n)$ it follows (using Parseval's formula, (4.16) and Lebesgue's convergence theorem) that

$$\langle u,g\rangle=\lim_{\varepsilon\to+0}\langle R_0(k^2+i\varepsilon)f,g\rangle=$$
$$=\lim_{\varepsilon\to+0}\int\limits_{\boldsymbol{R}^n}\frac{\hat{f}(\xi)\overline{\hat{g}(\xi)}}{|\xi|^2-k^2}\cdot\frac{|\xi|^2-k^2}{|\xi|^2-k^2-i\varepsilon}\,d\xi=\int\limits_{\boldsymbol{R}^n}\frac{\hat{f}(\xi)}{|\xi|^2-k^2}\overline{\hat{g}(\xi)}\,d\xi\,.$$

This proves (4.16)$'$, showing in particular that $\hat{u}\in L^1_{\text{loc}}(\boldsymbol{R}^n)$. Observing now that u verifies the conditions of Theorem 3.3, it follows from Theorem 3.3 that $u\in\mathcal{H}_{2,s}(\boldsymbol{R}^n)$ for all s. This proves the lemma.

We shall formulate now the limiting absorption principle for general Schrödinger operators.

THEOREM 4.2. *Let $H=-\Delta+V$ be a Schrödinger operator with potential V of class SR. Let $R(z)=(H-z)^{-1}$ be the resolvent of H. Consider $R(z)$ as an analytic operator valued function on $\boldsymbol{C}\setminus\sigma(H)$ with values in $B(L^{2,s},\mathcal{H}_{2,-s})$, for any $s>\frac{1}{2}$. Let $e_+(H)$ be the discrete set of positive eigenvalues of H, and let $\lambda\in\boldsymbol{R}_+\setminus e_+(H)$. Then, the following limits exist in the uniform operator topology of $B(L^{2,s},\mathcal{H}_{2,-s})$:*

$$\lim_{\substack{z\to\lambda\\ \pm\operatorname{Im}z>0}}R(z)=R^{\pm}(\lambda)\,. \tag{4.17}$$

Moreover, for any $f\in L^{2,s}(\boldsymbol{R}^n)$

$$R^{\pm}(\lambda)f=R_0^{\pm}(\lambda)f-R_0^{\pm}(\lambda)VR^{\pm}(\lambda)f\,. \tag{4.18}$$

In particular, $u^+=R^+(\lambda)f$ is a $\sqrt{\lambda}$-outgoing solution, and $u^-=R^-(\lambda)f$ is a $\sqrt{\lambda}$-incoming solution of the differential equation:

$$(-\Delta+V-\lambda)u=f\quad \text{in } \boldsymbol{R}^n\,. \tag{4.19}$$

In the following we shall refer to $R^{\pm}(\lambda)$ as the boundary values of $R(z)$ on the positive axis. Clearly the function $R^{\pm}(\lambda)$ (with values in $B(L^{2,s}, \mathcal{H}_{2,-s})$ for any $s > \frac{1}{2}$) is continuous on $\boldsymbol{R}_+ \setminus e_+(H)$ in the uniform operator topology of $B(L^{2,s}, \mathcal{H}_{2,-s})$.

Proof of Theorem 4.2. We shall prove the theorem for $R^+(\lambda)$, the proof for $R^-(\lambda)$ is similar. With no loss of generality we may assume that s is restricted to some interval $\frac{1}{2} < s \leqslant \frac{1}{2} + \varepsilon$, $\varepsilon > 0$. We shall choose $\varepsilon > 0$ sufficiently small so that the multiplication operator V is a compact operator from $\mathcal{H}_{2,-\frac{1}{2}-\varepsilon}(R^n)$ into $L^{2,\frac{1}{2}+\varepsilon}(R^n)$.

For any z in $\tilde{C}_+ = \boldsymbol{C}_+ \cup \boldsymbol{R}_+$ we define $T(z) \in B(\mathcal{H}_{2,-s}, \mathcal{H}_{2,-s})$ by

$$T(z)u = R_0^+(z)\,Vu \qquad \text{for } u \in \mathcal{H}_{2,-s}(\boldsymbol{R}^n)\,. \tag{4.20}$$

Here $R_0^+(z) = R_0(z) \in B(L^{2,s}, \mathcal{H}_{2,-s})$ if $\operatorname{Im} z > 0$, whereas $R_0^+(z)$ is defined by (4.1) if $z = \lambda \in \boldsymbol{R}_+$. From Theorem 4.1 and the compactness of V (considered as an operator from $\mathcal{H}_{2,-s}(\boldsymbol{R}^n)$ into $L^{2,s}(\boldsymbol{R}^n)$) it follows that $T(z)$ is a compact operator for every $z \in \tilde{C}_+$, and that, furthermore, the operator valued function $T(z)$ is continuous on $\tilde{C}_+$ in the uniform operator topology of $B(\mathcal{H}_{2,-s}, \mathcal{H}_{2,-s})$.

Consider the question of invertibility in $B(\mathcal{H}_{2,-s}, \mathcal{H}_{2,-s})$ of the operator $I + T(z)$ where I is the identity. We claim that $(I + T(z))^{-1}$ exists if and only if $z \in \tilde{C}_+ \setminus e_+(H)$. Indeed, suppose first that $\operatorname{Im} z > 0$. Using the resolvent equation:

$$R(z) + R_0(z)\,VR(z) = R_0(z)\,, \tag{4.21}$$

and (4.20), it follows that for any $f \in L^2(\boldsymbol{R}^n)$ and $u = R(z)f \in \mathcal{H}_2(\boldsymbol{R}^n)$, we have

$$(I + T(z))\,u = R_0(z)f\,. \tag{4.22}$$

Letting f vary on $L^2(\boldsymbol{R}^n)$, it follows from (4.22) that range $(I + T(z)) \supset \supset \mathcal{H}_2(\boldsymbol{R}^n)$, which implies that $\overline{\text{range}\,(I + T(z))} = \mathcal{H}_{2,-s}(\boldsymbol{R}^n)$. From this it follows by well known results on compact operators in a Hilbert space (the Fredholm-Riesz theory) that the inverse $(I+T(z))^{-1}$ exists in $B(\mathcal{H}_{2,-s}, \mathcal{H}_{2,-s})$.

Next, let $z = \lambda \in \boldsymbol{R}_+$. By the Fredholm-Riesz theory $I + T(\lambda)$ is invertible if and only if -1 is not an eigenvalue of $T(\lambda)$. Hence suppose that -1 is an eigenvalue of $T(\lambda)$ and let $u \in \mathcal{H}_{2,-s}(\boldsymbol{R}^n)$ be the corresponding eigenfunction. From (4.20) it follows that $u = -R_0^+(\lambda)(Vu)$, which implies that u is a $\sqrt{\lambda}$-outgoing solution of the differential equation: $(-\Delta + V)u = \lambda u$. Applying Lemma 4.2 it follows that $u \in \mathcal{D}(H)$, which in turn implies that λ is an eigenvalue of H. Conversely, let $\lambda > 0$ be an eigenvalue of H with a corresponding eigenfunction $u \in \mathcal{D}(H)$. Using (4.21) we find that $u + R_0(z)(Vu) = (\lambda - z)R_0(z)u$ for z in $\boldsymbol{C}_+$. Hence, letting $z \to \lambda$, using Theorem 3.3, we have: $u + R_0^+(\lambda)(Vu) = 0$, which shows that -1 is an eigenvalue of $T(\lambda)$. The above considerations show that $(I + T(z))^{-1}$ exists for z in $\tilde{C}_+$ if and only if $z \notin e_+(H)$.

Now, since $T(z)$ is continuous on $\tilde{C}_+$ in the uniform operator topology of $B(\mathcal{H}_{2,-s}, \mathcal{H}_{2,-s})$, it follows by elementary considerations that the operator valued function $(I + T(z))^{-1}$ is also continuous on $\tilde{\boldsymbol{C}}_+ \diagdown e_+(H)$ in the uniform operator topology of $B(\mathcal{H}_{2,-s}, \mathcal{H}_{2,-s})$. From (4.21) and (4.20) it follows that

$$R(z) = (I + T(z))^{-1}R_0(z) \quad \text{for } \operatorname{Im} z > 0 . \tag{4.23}$$

Using the continuity properties of $(I + T(z))^{-1}$ and $R_0(z)$ it follows that, for any $\lambda \in \boldsymbol{R}_+ \diagdown e_+(H)$,

$$\exists \lim_{\substack{z \to \lambda \\ \operatorname{Im} z > 0}} R(z) = (I + T(\lambda))^{-1} R_0^+(\lambda) \tag{4.24}$$

in the uniform operator topology of $B(L^{2,s}, \mathcal{H}_{2,-s})$. From the last result (or letting $z \to \lambda$ in (4.21)) we obtain (4.18). The other results mentioned in the theorem follow immediately from (4.18). This establishes the theorem.

We conclude this section with an approximation result which we shall need later on. It shows that boundary values $R^{\pm}(\lambda)$ of a Schrödinger operator with a SR potential depend continuously (in some sense) on V.

THEOREM 4.3. *Let $H = -\Delta + V$ be a Schrödinger operator with potential V of class SR. Let $H_j = -\Delta + V_j$, $j = 1, 2, \ldots$, be a sequence of Schrödinger operators with potentials V_j of class SR such that:* (i) $\lim V_j(x) = V(x)$ *for almost all x, and* (ii) $|V_j(x)| \leqslant W(x)$ *for all x and $j = 1, 2, \ldots$, where W is some function of class SR.*

Let $R(z)$ and $R_j(z)$ be the resolvents of H and H_j, respectively. Consider $R(z)$ and $R_j(z)$ as operator valued functions on $\boldsymbol{C} \diagdown \boldsymbol{R}$ with values in $B(L^{2,s}, \mathcal{H}_{2,-s})$, for some $s > \frac{1}{2}$. Denote by $R^{\pm}(\lambda)$ and $R_j^{\pm}(\lambda)$, $j = 1, 2, \ldots$, the boundary values of $R(z)$ and $R_j(z)$ on the positive axis, defined by Theorem 4.2. Let $\mathcal{K}$ be any compact set in $\boldsymbol{R}_+ \diagdown e_+(H)$. Then,

(i) *$\mathcal{K}$ does not contain any eigenvalues of H_j for $j \geqslant j_0$ sufficiently large.*

(ii) *The following limit relation holds in the operator topology of $B(L^{2,s}, \mathcal{H}_{2,-s})$:*

$$\lim_{j \to \infty} R_j^{\pm}(\lambda) = R^{\pm}(\lambda) , \tag{4.25}$$

uniformly for λ in $\mathcal{K}$.

PROOF. As was observed before, we may assume with no loss of generality that $\frac{1}{2} < s \leqslant \frac{1}{2} + \varepsilon$, $\varepsilon > 0$ arbitrary but fixed. We choose ε such that the multiplication operator $W: u \to W(x)u$, is a compact operator from $\mathcal{H}_{2,-\frac{1}{2}-\varepsilon}(\boldsymbol{R}^n)$ into $L^2(\boldsymbol{R}^n)$. We shall prove the theorem for $R^+(\lambda)$, the proof for $R^-(\lambda)$ is the same.

As in the proof of Theorem 4.2 we define for $\lambda \in \boldsymbol{R}_+$

$$T(\lambda) = R_0^+(\lambda)\,V\,, \qquad T_j(\lambda) = R_0^+(\lambda)\,V_j \quad \text{for } j = 1, 2, \dots, \tag{4.26}$$

where the multiplication operators V and V_j are considered as operators in $B(\mathcal{H}_{2,-s}, L^{2,s})$, and $R_0^+(\lambda) \in B(L^{2,s}, \mathcal{H}_{2,-s})$. Since by our assumption V and V_j are compact, it follows from (4.26) and Theorem 4.1 that $T(\lambda)$ and $T_j(\lambda)$ are compact operators in $B(\mathcal{H}_{2,-s}, \mathcal{H}_{2,-s})$ for every fixed λ, and that the operator valued functions $T(\lambda)$ and $T_j(\lambda)$ are continuous on $\boldsymbol{R}_+$ in the operator topology of $B(\mathcal{H}_{2,-s}, \mathcal{H}_{2,-s})$.

We claim that

$$\lim_{j\to\infty} T_j(\lambda) = T(\lambda) \tag{4.27}$$

in the operator topology of $B(\mathcal{H}_{2,-s}, \mathcal{H}_{2,-s})$, the convergence being uniform in λ on any compact subset of $\boldsymbol{R}_+$. Now, from (4.26) it is clear that (4.27) will follow if we show that

$$\lim_{j\to\infty} V_j = V \tag{4.28}$$

in the operator topology of $B(\mathcal{H}_{2,-s}, L^{2,s})$. To prove (4.28), set $U(x) = V(x)/W(x)$, $U_j(x) = V_j(x)/W(x)$, and note that $U_j(x) \to U(x)$ for almost all x, $|U_j(x)| \leqslant 1$ for all x, $j = 1, 2, \dots$. Consider the multiplication operators $U: f \to U(x)f$ and $U_j: f \to U_j(x)f$, as operators in $B(L^{2,s}, L^{2,s})$. It is clear that $s-\lim U_j = U$. This implies that if $\{f_j\}$ is a strongly convergent sequence in $L^{2,s}(\boldsymbol{R}^n)$ with $s-\lim f_j = f$, then $\exists s-\lim U_j f_j = Uf$ in $L^{2,s}(\boldsymbol{R}^n)$. This in turn implies, since $V = UW$, $V_j = U_j W$, and W is a compact operator in $B(\mathcal{H}_{2,-s}, L^{2,s})$, that is if $\{u_j\}$ is a weakly convergent sequence in $\mathcal{H}_{2,-s}(\boldsymbol{R}^n)$ with $w-\lim u_j = u$, then

$$\exists s-\lim V_j u_j = Vu \quad \text{in } L^{2,s}(\boldsymbol{R}^n)\,. \tag{4.28'}$$

By an obvious argument used already before [13], this last property implies (4.28) and establishes (4.27).

[13] See end of proof of Theorem 4.1.

Now, from the proof of Theorem 4.2 it follows that the inverse $(I+T(\lambda))^{-1}$ $\big((I+T_j(\lambda))^{-1}\big)$ exists if and only if $\lambda\in\boldsymbol{R}_+\diagdown e_+(H)$, $\big(\lambda\in\boldsymbol{R}_+\diagdown e_+(H_j)\big)$. This and (4.27) imply that if $\mathcal{K}$ is a compact set in $\boldsymbol{R}_+\diagdown e_+(H)$, then $\mathcal{K}$ does not contain any eigenvalues of H_j for all j sufficiently large, and that the following limit relation holds in the operator topology of $B(\mathcal{H}_{2,-s},\mathcal{H}_{2,-s})$:

$$\lim_{j\to\infty}(I+T_j(\lambda))^{-1}=(I+T(\lambda))^{-1}, \tag{4.29}$$

uniformly in λ on $\mathcal{K}$. Noting that (by (4.23)) we have

$$R^+(\lambda)=(I+T(\lambda))^{-1}R_0^+(\lambda)\,,\qquad R_j^+(\lambda)=(I+T_j(\lambda))^{-1}R_0^+(\lambda)\,; \tag{4.30}$$

it follows from (4.29), (4.30) and the continuity of $R_0^+(\lambda)$, that (4.25) holds. This completes the proof of the theorem.

5. – The generalized eigenfunctions.

The question whether there exists a good eigenfunction expansion theorem for the Schrödinger operator $H=-\Delta+V$ is related to the question of existence of a « good » family of generalized eigenfunctions which behave like plane waves. This family is a function $\phi(x,\xi)$, defined for $x\in\boldsymbol{R}^n$ and $\xi\in\boldsymbol{R}^n\diagdown\{0\}$ $\big(|\xi|^2\notin e_+(H)\big)$, which satisfies the differential equation

$$(-\Delta_x+V(x)-|\xi|^2)\,\phi(x,\xi)=0\,, \tag{5.1}$$

and which has the form: $\phi(x,\xi)=\exp[ix\cdot\xi]+v(x,\xi)$ with $v\to0$ as $|x|\to\infty$ (in some sense). In particular v should satisfy the equation

$$(-\Delta_x+V(x)-|\xi|^2)\,v(x,\xi)=-V(x)\exp[ix\cdot\xi]\quad\text{ in } \boldsymbol{R}^n\,. \tag{5.1$'$}$$

Now, if V is of class SR and if, moreover, $V\in L^{2,s}(\boldsymbol{R}^n)$ for some $s>\frac12$, then the limiting absorption principle (Theorem 4.2) yields two solutions of (5.1′). Hence we find in this case two families of generalized eigenfunctions given by

$$\phi_\pm(x,\xi)=\exp[ix\cdot\xi]-R^\mp\big(|\xi|^2\big)\big[V(\cdot)\exp[i(\cdot,\xi)]\big](x)\,. \tag{5.2}$$

The eigenfunction expansion theorem that one obtains with the aid of the generalized eigenfunctions $\phi_\pm(x,\xi)$ defined by (5.2) is applicable to Schrödinger operators with potentials having (roughly) a decay rate

$$V(x)=O\big(|x|^{-(n+1)/2-\varepsilon}\big)\quad\text{ as } |x|\to\infty\,. \tag{5.3}$$

Condition (5.3) is (essentially) the rate of decay assumption on V imposed in most works on the eigenfunction expansion theorem. (We disregard here weaker integral conditions on V which *on the average* imply (5.3).) Now, the rate of decay condition on V, given by (5.3), depends on the space dimension n. The question arises whether such a strong condition is really necessary for the existence of the two families of generalized eigenfunctions $\phi_\pm(x, \xi)$. In the following we shall show that condition (5.3) is indeed not necessary and that the appropriate families of generalized eigenfunctions $\phi_\pm(x, \xi)$ exist under the weaker assumption:

$$V(x) = O(|x|^{-1-\varepsilon}) \qquad \text{as } |x| \to \infty,$$

or to that matter under the still weaker assumption that condition (1.3) holds. It should be remarked, however, that in the general case the functions $\phi_\pm(x, \xi)$ need not be continuous functions in *both* variables x and ξ (as they are if (5.3) holds), but only be continuous in x and belong to a certain class of measurable functions in x and ξ.

In order to prove the existence of the generalized eigenfunctions under minimal decay assumptions on the potential we shall use, in addition to the limiting absorption principle, two lemmas. In connection with the first lemma, and also for later use, we introduce certain classes of continuous functions on $\boldsymbol{R}^n$. We denote by $C^{0,s}(\boldsymbol{R}^n)$, s any real number, the class of continuous functions $u(x)$ on $\boldsymbol{R}^n$ such that $(1+|x|)^s u(x) \in L^\infty(\boldsymbol{R}^n)$. We consider $C^{0,s}(\boldsymbol{R}^n)$ as a Banach space with norm

$$(5.4) \qquad |||u|||_{0,s} = \sup_{\boldsymbol{R}^n} (1+|x|)^s\, |u(x)|\,.$$

For any $0<\theta<1$ and real s we denote by $C^{\theta,s}(\boldsymbol{R}^n)$ the subclass of functions u in $C^{0,s}(\boldsymbol{R}^n)$ such that $(1+|x|)^s u(x)$ verifies on $\boldsymbol{R}^n$ a uniform Hölder condition of order θ. We shall consider $C^{\theta,s}(\boldsymbol{R}^n)$ as a Banach space with norm

$$(5.5) \qquad |||u|||_{\theta,s} = |||u|||_{0,s} + \sup_{\substack{x,y\\ 0<|x-y|<1}} \left[(1+|x|)^s \frac{|u(x)-u(y)|}{|x-y|^\theta}\right].$$

The following lemma combines a regularity result with an a-priori estimate for solutions of certain Schrödinger equations.

LEMMA 5.1. *Let $u(x)$ be a function in $L^{2,s}(\boldsymbol{R}^n) \cap \mathcal{H}_2^{\mathrm{loc}}(\boldsymbol{R}^n)$ for some real s. Suppose that u satisfies the differential equation*

$$-\Delta u + qu = f \qquad \textit{in } \boldsymbol{R}^n$$

where $q(x) \in L^2_{\mathrm{loc}}(\boldsymbol{R}^n)$ and $f(x) \in L^{2,s}(\boldsymbol{R}^n)$.

Suppose also that there exist a number θ, $0<\theta<\frac{1}{2}$, and positive constants Q and F such that

$$\sup_{x} \int_{|y-x|\leqslant 1} |q(y)|^2 \, |y-x|^{-n-2\theta+4} \, dy \leqslant Q^2 \tag{5.6}$$

and

$$\sup_{x} \, (1+|x|)^{2s} \int_{|y-x|\leqslant 1} |f(y)|^2 \, |y-x|^{-n-2\theta+4} \, dy \leqslant F^2 \, . \tag{5.7}$$

Then $u \in C^{\theta,s}(\boldsymbol{R}^n) \cap \mathcal{H}_{2,s}(\boldsymbol{R}^n)$, and the following estimate holds:

$$|||u|||_{\theta,s} + \|u\|_{2,s} \leqslant \gamma (Q+1)^{\nu} (\|u\|_{0,s} + F) \, , \tag{5.8}$$

where ν is a constant depending only on θ and n; γ is a constant depending only on θ, n and s.

The proof of Lemma 5.1 is given in Appendix C.

The second result we shall need is

LEMMA 5.2. *Let Γ be a C^∞ compact $n-1$ dimensional manifold imbedded in $\boldsymbol{R}^n$. Let $d\sigma$ be the measure induced on Γ by the Lebesgue measure, and let $L^2(\Gamma)$ be the class of L^2 functions on Γ with respect to the measure $d\sigma$. With any given function $g \in L^2(\Gamma)$ associate a function $g(x)$ on $\boldsymbol{R}^n$, defined by*

$$\tilde{g}(x) = \int_{\Gamma} g(\xi) \exp[-i\xi\cdot x] \, d\sigma_\xi \, . \tag{5.9}$$

Then, $\tilde{g}(x) \in \mathcal{H}_{m,-s}(\boldsymbol{R}^n) \cap C^\infty(\boldsymbol{R}^n)$ for any $m \geqslant 0$ and any $s > \frac{1}{2}$. The following estimate holds:

$$\|\tilde{g}\|_{m,-s} \leqslant C_{m,s} \|g\|_{L^2(\Gamma)} \, , \tag{5.10}$$

where $C_{m,s}$ is a constant depending only on m, s and Γ.

This useful lemma is due to *Y. Kannai* [14]. It was observed by S. T. *Kuroda* that the lemma follows easily, by duality, from the trace theorem (see end of section 2). The following proof follows *Kuroda*'s observation.

[14] Oral communication.

PROOF OF LEMMA. Let $\varphi \in C_0^\infty(\boldsymbol{R}^n)$. It follows from (5.9) that

$$\int_{\boldsymbol{R}^n} \varphi(x)\tilde{g}(x)\,dx = \int_{\boldsymbol{R}^n} \varphi(x)\Big(\int_{\Gamma} g(\xi)\exp[-i\xi\cdot x]\,d\sigma\Big)\,dx = (2\pi)^{n/2}\int_{\Gamma}\hat{\varphi}(\xi)\,g(\xi)\,d\sigma\,. \tag{5.11}$$

By the trace theorem (applied to $\hat{\varphi}$) and the definition of the s-norm, we have

$$\|\tau\hat{\varphi}\|_{L^2(\Gamma)} \leqslant C\|\hat{\varphi}\|_s = C\|\varphi\|_{0,s}\,, \tag{5.12}$$

where $\tau\hat{\varphi}$ denotes the trace of $\hat{\varphi}$ on Γ, and C is a constant depending only on s and Γ. Hence, combining (5.11) and (5.12), we get

$$\Big|\int_{\boldsymbol{R}^n} \varphi(x)\tilde{g}(x)\,dx\Big| \leqslant (2\pi)^{n/2}\|\tau\varphi\|_{L^2(\Gamma)}\|g_{L^2(\Gamma)}\| \leqslant (2\pi)^{n/2}\,C\|\varphi\|_{0,s}\|g\|_{L^2(\Gamma)} \quad \text{for } \forall\varphi \in C_0^\infty(\boldsymbol{R}^n)\,,$$

which implies that $\tilde{g} \in L^{2,-s}(\boldsymbol{R}^n)$ and that

$$\|\tilde{g}\|_{0,-s} \leqslant (2\pi)^{n/2}\,C\|g\|_{L^2(\Gamma)}\,. \tag{5.13}$$

This proves (5.10) for $m = 0$. The estimate (5.10) for general m follows from (5.13) by differentiation, observing that $D^\alpha\tilde{g} = \tilde{g}^\alpha$ where $g^\alpha(\xi) = (-\xi)^\alpha g(\xi)$.

REMARK. Set $\tilde{g}_k(x) = \tilde{g}(kx)$ for any $k > 0$, $\tilde{g}(x)$ defined by (5.9). It follows from the lemma that for any $s > \frac{1}{2}$, any integer $m \geqslant 0$, and any compact interval $\mathcal{K} \subset \boldsymbol{R}_+$, we have

$$\|\tilde{g}_k\|_{m,-s} \leqslant C\|g\|_{L^2(\Gamma)} \quad \text{for } \forall g \in L^2(\Gamma) \tag{5.14}$$

and $\forall k \in \mathcal{K}$, where C is some constant.

We turn now to the construction of the two families of generalized eigenfunctions $\phi_\pm(x, \xi)$ for the Schrödinger operator $-\Delta + V$. As was already mentioned we shall assume that V verifies condition (1.3), or, what amounts to the same thing, we shall assume that for some $\varepsilon > 0$ and $0 < \theta < \frac{1}{2}$, the following condition holds:

$$\sup_{x\in\boldsymbol{R}^n}\Big[(1+|x|)^{2+2\varepsilon}\int_{|y-x|\leqslant 1}|V(y)|^2\,|y-x|^{-n+4-2\theta}\,dy\Big] < \infty\,. \tag{5.15}$$

It was observed before that a function V which verifies condition (5.15) is of class SR. We shall also use the following notation. With the set of positive eigenvalues $e_+(H)$ we associate the set in $\boldsymbol{R}_+$:

$$e_+(H)^{\frac{1}{2}} = \{k\colon k \in \boldsymbol{R}_+,\ k^2 \in e_+(H)\}\,, \tag{5.16}$$

and the set in $\boldsymbol{R}^n$:

$$\mathcal{N}(H) = \{\xi\colon \xi \in \boldsymbol{R}^n,\ |\xi|^2 \in e_+(H)\} \cup \{0\}\,. \tag{5.16$'$}$$

We shall use polar coordinates k, ω in the $\boldsymbol{R}^n_\xi$ space ($k = |\xi|$, $\omega = \xi/|\xi|$). We shall denote by Σ the unit $n-1$ sphere in $\boldsymbol{R}^n$, $\Sigma = \{\xi\colon |\xi| = 1\}$. We shall denote by $d\omega$ the measure on Σ induced by the Lebesgue measure on $\boldsymbol{R}^n$.

THEOREM 5.1. *Let $H = -\Delta + V$ be a Schrödinger operator with potential V satisfying condition* (5.15). *There exist two families $\phi_\pm(x, \xi)$ of generalized eigenfunctions of H, defined for every $\xi \in \boldsymbol{R}^n \setminus \mathcal{N}(H)$, having the following properties:*

(i) *As a function of x and ξ, $\phi_\pm(x, \xi)$ is a measurable function of class $L^2_{\mathrm{loc}}(\boldsymbol{R}^n \times (\boldsymbol{R}^n \setminus \mathcal{N}))$.*

(ii) *For every fixed ξ the function $\phi_\pm(x, \xi)$ belongs to $C(\boldsymbol{R}^n_x) \cap \mathcal{H}^{\mathrm{loc}}_2(\boldsymbol{R}^n_x)$ and satisfies the differential equation* (5.1).

(iii) *Introduce polar coordinates and write: $\psi_\pm(x, k, \omega) = \phi_\pm(x, k\omega)$. Then for fixed $(x, k) \in \boldsymbol{R}^n \times (\boldsymbol{R}_+ \setminus e_+(H)^{\frac{1}{2}})$ the function $\psi_\pm(x, k, \omega)$ belongs to $L^2(\Sigma)$. Moreover, the vector valued function $\psi_\pm(x, k, \cdot)$, with values in $L^2(\Sigma)$, is a continuous function of x and k on $\boldsymbol{R}^n \times (\boldsymbol{R}_+ \setminus e_+(H)^{\frac{1}{2}})$.*

(iv) *For any function g in $L^2(\Sigma)$, define*

$$\phi^g_\pm(x, k) = \int_\Sigma \phi_\pm(x, k\omega)\, g(\omega)\, d\omega\,, \qquad \phi^g_0(x, k) = \int_\Sigma \exp[ik\omega\cdot x]\, g(\omega)\, d\omega\,. \tag{5.17}$$

Then, for a fixed $k \in \boldsymbol{R}_+ \setminus e_+(H)^{\frac{1}{2}}$, the function $\phi^g_\pm(x, k)$ has the representation:

$$\phi^g_\pm(x, k) = \phi^g_0(x, k) - R^\mp(k^2)[V(\cdot)\phi^g_0(\cdot, k)](x)\,, \tag{5.18}$$

where $R^\mp(k^2)$ are the boundary values of the resolvent of H defined by (4.17).

In particular $\phi^g_{\pm}(x, k)$ *lies in* $\mathcal{H}_{2,-s}(\mathbf{R}^n) \cap C^{0,-s}(\mathbf{R}^n)$, *for any* $s > \frac{1}{2}$, *and satisfies the differential equation*

$$(5.19) \qquad (-\Delta_x + V(x) - k^2)\phi^g_{\pm}(x, k) = 0 .$$

REMARK 1. Our proof will show that the functions $\phi_{\pm}(x, \xi)$, for any fixed ξ, are in $C^{\theta,-r}(\mathbf{R}^n_x) \cap \mathcal{H}_{2,-r}(\mathbf{R}^n_x)$ for some $\theta > 0$ (the exponent in condition (5.15)) and any $r > (n+1)/2$.

REMARK 2. The families $\phi_{\pm}(x, \xi)$ are (essentially) uniquely defined by the conditions of the theorem. Indeed, suppose that $\tilde{\phi}_{\pm}(x, \xi)$ is a second pair of families of generalized eigenfunctions which also verify the conditions of Theorem 5.1. It follows readily (using (5.18)) that

$$\int\limits_{\mathbf{R}^n \times \mathbf{R}^n}\!\!\int \phi_{\pm}(x, \xi) f(x) g(\xi)\, dx\, d\xi = \int\limits_{\mathbf{R}^n \times \mathbf{R}^n}\!\!\int \tilde{\phi}_{\pm}(x, \xi) f(x) g(\xi)\, dx\, d\xi$$

for any $f \in C_0(\mathbf{R}^n)$ and $g \in C_0(\mathbf{R}^n \setminus \mathcal{N})$ [15].

This implies that $\phi_{\pm}(x, \xi) = \tilde{\phi}_{\pm}(x, \xi)$ for almost all (x, ξ) in $\mathbf{R}^n \times (\mathbf{R}^n \setminus \mathcal{N})$. From the continuity of $\phi_{\pm}$ and $\tilde{\phi}_{\pm}$ in x, it follows further that for almost all ξ we have that $\phi_{\pm}(x, \xi) = \tilde{\phi}_{\pm}(x, \xi)$ for all x.

PROOF OF THEOREM 5.1. We shall prove the theorem for the family $\phi_-(x, \xi)$, the proof for $\phi_+(x, \xi)$ is similar.

We start by approximating V by potentials which decay sufficiently rapidly at infinity. More precisely, we choose a sequence of real functions $\{V_j(x)\}_{j=1}^{\infty}$ having the following properties. (i) $V_j(x) \in L^{2,s_0}(\mathbf{R}^n)$ for some $s_0 > \frac{1}{2}$ and all j. (ii) $V_j(x) \to V(x)$ for almost all x in $\mathbf{R}^n$. (iii) $|V_j(x)| \leqslant W(x)$ for all x in $\mathbf{R}^n$ and all j, where $W(x)$ is a function satisfying condition (5.15) for some $\varepsilon > 0$ and $0 < \theta < \frac{1}{2}$. (Thus we may always take for $\{V_j\}$ the sequence defined by: $V_j(x) = V(x)$ for $|x| \leqslant j$, $V_j(x) \equiv 0$ for $|x| > j$. In case $V \in L^{2,s_0}(\mathbf{R}^n)$ for some $s_0 > \frac{1}{2}$, we may take $V_j = V$ for all j.) With the sequence $\{V_j\}$ we associate the sequence of Schrödinger operators $H_j = -\Delta + V_j$, $j = 1, 2, \ldots$. We shall denote by $R_j^+(\lambda)$ the boundary values of the resolvent of H_j on the upper edge of the positive axis, as defined by Theorem 4.2. Thus, $R_j^+(\lambda)$ is a continuous operator valued function, defined for $\lambda \in \mathbf{R}_+ \setminus e_+(H_j)$, with values in $B(L^{2,s}, \mathcal{H}_{2,-s})$ for any $s > \frac{1}{2}$. In the following s will denote an arbitrary but *fixed* number such that $\frac{1}{2} < s < \min(s_0, \frac{1}{2} + \delta)$ where $\delta > 0$ is chosen sufficiently small so that the function $W_1(x) = (1 + |x|)^{\delta} W(x)$ verifies also condition (5.15).

[15] $C_0(\Omega)$ denotes the class of continuous functions with a compact support in Ω.

Write $\phi_0(x, \xi) = \exp(ix\cdot\xi)$ and denote by $\mathcal{N}_j$ the set $\mathcal{N}(H_j)$ defined by (5.16′). For any $\xi \in \boldsymbol{R}^n \setminus \mathcal{N}_j$, define

$$\phi_j(x, \xi) = \phi_0(x, \xi) - R_j^+(|\xi|^2)[V_j(\cdot)\phi_0(\cdot, \xi)](x) . \tag{5.20}$$

(Note that $\phi_j(x, \xi)$ is well defined since $V_j(x)\phi_0(x, \xi) \in L^{2,s}(\boldsymbol{R}_x^n)$.)

From (5.20) it follows that $\phi_j(x, \xi) \in \mathcal{H}_{2,-r}(\boldsymbol{R}_x^n)$ for any $r > n/2$, and that ϕ_j satisfies the differential equation

$$(-\Delta_x + V_j(x) - |\xi|^2)\phi_j(x, \xi) = 0 . \tag{5.21}$$

Applying Lemma 5.1 to $\phi_j(x, \xi)$ it follows further that $\phi_j(x, \xi)$ is a Hölder continuous function of x. More precisely, it follows from the lemma that

$$|||\phi_j(\cdot, \xi)|||_{\theta,-r} \leqslant C_\xi \|\phi_j(\cdot, \xi)\|_{0,-r} \tag{5.22}$$

where C_ξ is a locally bounded function of ξ on $\boldsymbol{R}^n$. From (5.20) it follows that $\phi_j(\cdot, \xi)$ is a continuous function of ξ with values in $L^{2,-r}(\boldsymbol{R}^n)$. This and (5.22) imply that $\phi_j(x, \xi)$ is a locally bounded function of x and ξ. Finally, applying Lemma 5.1 to the function $u(x) = \phi_j(x, \xi) - \phi_j(x, \xi')$, which satisfies the differential equation

$$(-\Delta + V_j - |\xi|^2)u(x) = (|\xi|^2 - |\xi'|^2)\varphi_j(x, \xi) ,$$

it follows easily (with the aid of the observations just made) that $\phi_j(x, \xi)$ is a continuous function of x and ξ on $\boldsymbol{R}^n \times (\boldsymbol{R}^n / \mathcal{N}_j)$.

Let $\mathcal{K}$ be any compact set in $\boldsymbol{R}_+ \setminus e_+(H)$ ($\mathcal{K}$ will be fixed throughout our discussion). Since the sequence of operators $\{H_j\}$ verifies the conditions of Theorem 4.3, it follows that there exists a positive integer $j_0 = j_0(\mathcal{K})$ such that H_j has no eigenvalues in $\mathcal{K}$ for $\forall j \geqslant j_0$. For any $k > 0$ with $k^2 \in \mathcal{K}$, $j \geqslant j_0(\mathcal{K})$ and $g \in L^2(\Sigma)$, define

$$\phi_j^g(x, k) = \phi_0^g(x, k) - R_j^+(k^2)[V_j(\cdot)\phi_0^g(\cdot, k)](x) , \tag{5.23}$$

where as in (5.17) we let

$$\phi_0^g(x, k) = \int_\Sigma \phi_0(x, k\omega) g(\omega)\, d\omega .$$

Now, be Lemma 5.2 the function $\phi_0^g(x, k)$ belongs to $\mathcal{H}_{2,-s}(\boldsymbol{R}_x^n)$. This implies that, for a fixed k, $\phi_j^g(x, k)$ is a well defined function in $\mathcal{H}_{2,-s}(\boldsymbol{R}_x^n)$. Since $\phi_j^g(x, k)$ is a solution of the differential equation (5.21) (with $|\xi| = k$)

it follows further, upon application of Lemma 5.1, that $\phi_j^g(x, k)$ is a continuous function on $\boldsymbol{R}^n$ of class $C^{\theta,-s}(\boldsymbol{R}_x^n)$. (Throughout the proof $0 < \theta < \frac{1}{2}$ stands for the θ-exponent in condition (5.15) for the function W.)

We propose to show that the sequence of vector valued functions $\phi_j^g(\cdot\,, k)$, with values in $C^{0,-s}(\boldsymbol{R}^n) \cap \mathcal{H}_{2,-s}(\boldsymbol{R}^n)$, converges to a limit, uniformly with respect to k^2 in $\mathcal{K}$ and g in the unit ball of $L^2(\Sigma)$. More precisely, we are going to show that there exist a constant $C = C(\mathcal{K})$ and a sequence of positive numbers $\varepsilon_j = \varepsilon_j(\mathcal{K})$, with $\varepsilon_j \to 0$, such that

$$\left.\begin{matrix} \|\phi_j^g(\cdot\,, k)\|_{2,-s} \\ |||\phi_j^g(\cdot\,, k)|||_{\theta,-s} \end{matrix}\right\} \leqslant C\|g\|_{L^2(\Sigma)}\,, \tag{5.24}$$

and

$$\left.\begin{matrix} \|\phi_m^g(\cdot\,, k) - \phi_j^g(\cdot\,, k)\|_{2,-s} \\ |||\phi_m^g(\cdot\,, k) - \phi_j^g(\cdot\,, k)|||_{0,-s} \end{matrix}\right\} \leqslant \varepsilon_j\|g\|_{L^2(\Sigma)}\,, \tag{5.24'}$$

for $\forall g \in L^2(\Sigma)$, $\forall k^2 \in \mathcal{K}$ and all integers m, j such that $m > j \geqslant j_0(\mathcal{K})$.

We shall first prove the L^2 estimates part of (5.24) and (5.24′). To this end consider the operators

$$\tilde{T}(\lambda) = R^+(\lambda)\,V\,, \qquad \tilde{T}_j(\lambda) = R_j^+(\lambda)V_j\,, \tag{5.25}$$

defined for any $\lambda \in \mathcal{K}$ and $j \geqslant j_0(\mathcal{K})$, where the multiplication operators V and V_j are taken as compact operators from $\mathcal{H}_{2,-s}(\boldsymbol{R}^n)$ into $L^{2,s}(\boldsymbol{R}^n)$. Recalling the properties of the boundary operators $R^+(\lambda)$ and $R_j^+(\lambda)$, it follows that $\tilde{T}(\lambda)$ and $\tilde{T}_j(\lambda)$ are continuous functions of λ with values in $B(\mathcal{H}_{2,-s}, \mathcal{H}_{2,-s})$. By Theorem 4.3, we have

$$\lim_{j\to\infty} R_j^+(\lambda) = R^+(\lambda) \qquad \text{in } B(L^{2,s}, \mathcal{H}_{2,-s})\,, \tag{5.26}$$

uniformly for λ in $\mathcal{K}$. Also, in the process of proving Theorem 4.3 we have shown (see (4.28)) that

$$\lim_{j\to\infty} V_j = V \qquad \text{in } B(\mathcal{H}_{2,-s}, L^{2,s}) \tag{5.26'}$$

Hence, combining (5.25), (5.26) and (5.26′), it follows that

$$\lim_{j\to\infty} \tilde{T}_j(\lambda) = \tilde{T}(\lambda) \tag{5.27}$$

in the operator topology of $B(\mathcal{H}_{2,-s}, \mathcal{H}_{2,-s})$, the convergence being uniform on $\mathcal{K}$. Now, it follows from (5.23) and (5.25) that

$$\phi_j^g(x, k) = \phi_0^g(x, k) - \tilde{T}_j(k^2)\phi_0^g(\cdot, k)(x). \tag{5.28}$$

By Lemma 5.2 there exists a constant γ_0 such that

$$\|\phi_0^g(\cdot, k)\|_{2,-s} \leqslant \gamma_0 \|g\|_{L^2(\Sigma)} \tag{5.29}$$

for $\forall g \in L^2(\Sigma)$ and $\forall k^2 \in \mathcal{K}$. Combining (5.28), (5.29) and (5.27), we conclude that

$$\begin{aligned}\|\phi_j^g(\cdot, k)\|_{2,-s} &\leqslant \gamma_0(1 + \|\tilde{T}_j(k^2)\|_{B(\mathcal{H}_{2,-s},\mathcal{H}_{2,-s})})\|g\|_{L^2(\Sigma)} \\ &\leqslant C_0\|g\|_{L^2(\Sigma)},\end{aligned} \tag{5.30}$$

and also that

$$\begin{aligned}\|\phi_m^g(\cdot, k) - \phi_j^g(\cdot, k)\|_{2,-s} &\leqslant \gamma_0\|\tilde{T}_m(k^2) - \tilde{T}_j(k^2)\|_{B(\mathcal{H}_{2,-s},\mathcal{H}_{2,-s})}\|g\|_{L^2(\Sigma)} \\ &\leqslant \tilde{\varepsilon}_j\|g\|_{L^2(\Sigma)},\end{aligned} \tag{5.30$'$}$$

for $\forall m > j \geqslant j_0(\mathcal{K})$, $\forall k^2 \in \mathcal{K}$ and $\forall g \in L^2(\Sigma)$. Here C_0 is some constant and $\{\tilde{\varepsilon}_j\}$ is a sequence of positive numbers such that $\tilde{\varepsilon}_j \to 0$ as $j \to \infty$. This yields the L^2 estimates part of (5.24) and (5.24′).

To derive the second estimate of (5.24), we use once more the fact that $\phi_j^g(x, k)$ is a solution of the differential equation:

$$(-\Delta_x + V_j - k^2)\phi_j^g(x, k) = 0,$$

where $|V_j| \leqslant W$ for $\forall j$, W satisfying (5.15). Applying Lemma 5.1 to the function $\phi_j^g(x, k)$ it follows that there exists a constant $\gamma_1 = \gamma_1(\mathcal{K})$ such that

$$|||\phi_j^g(\cdot, k)|||_{\theta,-s} \leqslant \gamma_1\|\phi_j^g(\cdot, k)\|_{0,-s} \tag{5.31}$$

for $\forall j \geqslant j_0$, $k^2 \in \mathcal{K}$ and $\forall g \in L^2(\Sigma)$. Combining (5.31) and (5.30), the second estimate (5.24) follows.

Finally, to prove the second estimate in (5.24′), we set

$$u_{jm}^g(x, k) = \phi_m^g(x, k) - \phi_j^g(x, k), \quad m > j \geqslant j_0. \tag{5.32}$$

For any $x^0 \in \boldsymbol{R}^n$ and $0 < \varrho \leqslant 1$, we have

$$\begin{aligned}\Big(\int_{|x-x^0|<\varrho} |u_{jm}^g(x^0, k)|^2 dx\Big)^{\frac{1}{2}} &\leqslant \Big(\int_{|x-x^0|<\varrho} |u_{jm}^g(x, k)|^2 dx\Big)^{\frac{1}{2}} \\ &\quad + \Big(\int_{|x-x^0|<\varrho} |u_{jm}^g(x, k) - u_{jm}^g(x_0, k)|^2 dx\Big)^{\frac{1}{2}}.\end{aligned} \tag{5.33}$$

Now, using the inequality

$$\Big(\int_{|x-x^0|<\varrho} |u^g_{jm}(x, k)|^2\, dx\Big)^{\frac12} \leqslant 2^{s/2}(|x^0|+1)^s \|u^g_{jm}(\cdot, k)\|_{0,-s},$$

it follows from (5.30′) and (5.32) that

$$(5.34)\qquad \Big(\int_{|x-x^0|<\varrho} |u^g_{jm}(x, k)|^2\, dx\Big)^{\frac12} \leqslant 2^{s/2}(|x^0|+1)^s \|u^g_{jm}(\cdot, k)\|_{0,-s} \leqslant 2^{s/2}\tilde{\varepsilon}_j(|x^0|+1)^s\|g\|_{L^2(\Sigma)}.$$

Using the second inequality in (5.24), taking note of the definition of the weighted Hölder norm (5.5), we have

$$\begin{aligned}|u^g_{jm}(x, k) - u^g_{jm}(x^0, k)| &\leqslant |\phi^g_j(x, k) - \phi^g_j(x^0, k)| + |\phi^g_m(x, k) - \phi^g_m(x^0, k)| \\ &\leqslant |x-x^0|^\theta(1+|x^0|)^s(|||\phi^g_j(\cdot, k)|||_{\theta,-s} + |||\phi^g_m(\cdot, k)|||_{\theta,-s}) \\ &\leqslant 2C|x-x^0|^\theta(1+|x^0|)^s\|g\|_{L^2(\Sigma)}.\end{aligned}$$

Hence, by integration,

$$(5.35)\qquad \Big(\int_{|x-x^0|<\varrho} |u^g_{jm}(x, k) - u^g_{jm}(x^0, k)|^2\, dx\Big)^{\frac12} \leqslant C_1\varrho^{n/2+\theta}(1+|x^0|)^s\|g\|_{L^2(\Sigma)}$$

where C_1 is a constant which does not depend on x^0, $k^2\in\mathcal{K}$, or on g.

Combining now (5.33), (5.34) and (5.35), we find that

$$(5.36)\qquad \Omega_n^{\frac12}|u^g_{jm}(x^0, k)| \leqslant 2^{s+1}(\tilde{\varepsilon}_j\varrho^{-n/2} + C_1\varrho^{\theta/2})(|x^0|+1)^s\|g\|_{L^2(\Sigma)},$$

for any $\varrho\leqslant 1$, Ω_n denoting the volume of the unit ball in $\boldsymbol{R}^n$. Choosing $\varrho = \varepsilon_j^{2/(n+\theta)}$ if $\varepsilon_j\leqslant 1$ and $\varrho = 1$ if $\varepsilon_j > 1$, it follows from (5.36) that

$$|\phi^g_m(x^0, k) - \phi^g_j(x^0, k)| \leqslant \varepsilon'_j(|x^0|+1)^s\|g\|_{L^2(\Sigma)},$$

for $\forall x^0\in\boldsymbol{R}^n$, $\forall k^2\in\mathcal{K}$, $\forall g\in L^2(\Sigma)$ and all integers $m > j \geqslant j_0$, where $\varepsilon'_j = \text{const}\cdot\tilde{\varepsilon}_j^{\theta/(n+\theta)} \to 0$ as $j\to\infty$. This establishes the last estimate in (5.24′).

Returning to the functions $\phi_j(x, \xi)$ defined by (5.20), it is easy to see that the previous considerations imply the following

Spherical Mean Estimates for $\phi_j(x, k\omega)$. Let $\mathcal{K}$ be compact set in $\boldsymbol{R}_+\setminus e_+(H)$. Then there exist a constant C and a sequence of positive numbers $\{\varepsilon_j\}$, $\varepsilon_j\to 0$, such that

$$(5.37)\qquad \Big(\int_\Sigma |\phi_j(x, k\omega|^2\Big)^{\frac12} \leqslant C(1+|x|)^s,$$

and

$$(5.37')\qquad \Big(\int_\Sigma |\phi_m(x, k\omega) - \phi_j(x, k\omega)|^2\, d\omega\Big)^{\frac{1}{2}} \leqslant \varepsilon_j(1 + |x|)^s$$

for $\forall x \in \boldsymbol{R}^n$, $\forall k^2 \in \mathcal{K}$ *and all integers* $m > j \geqslant j_0(\mathcal{K})$.

Indeed, to obtain these estimates observe first that (5.20) and (5.23) yield the following relation:

$$(5.38)\qquad \phi_j^g(x, k) = \int_\Sigma \phi_j(x, k\omega)\, g(\omega)\, d\omega\,,$$

where (5.38) holds pointwisely (since ϕ_j^g and ϕ_j are continuous in x and k). From (5.38) and (5.4) it follows that for every fixed x and $k^2 \in \mathcal{K}$

$$(5.39)\qquad \Big(\int_\Sigma |\phi_j(x, k\omega)|^2 d\omega\Big)^{\frac{1}{2}} = \sup_{\substack{g\in L^2(\Sigma)\\ \|g\|=1}} |\phi_j^g(x, k)| \leqslant (1 + |x|)^s \sup_{\substack{g\in L^2(\Sigma)\\ \|g\|=1}} |||\phi_j^g(\cdot, k)|||_{0,-s}\,,$$

and

$$(5.39')\qquad \Big(\int_\Sigma |\phi_m(x, k\omega) - \phi_j(x, k\omega)|^2\, d\omega\Big)^{\frac{1}{2}} = \sup_{\substack{g\in L^2(\Sigma)\\ \|g\|=1}} |\phi_m^g(x, k) - \phi_j^g(x, k)| \leqslant$$
$$\leqslant (1 + |x|)^s \sup_{\substack{g\in L^2(\Sigma)\\ \|g\|=1}} |||\phi_m^g(\cdot, k) - \phi_j^g(\cdot, k)|||_{0,-s}\,.$$

Hence, combining (5.39) and (5.24) we obtain the estimate (5.37). Similarly, combining (5.39′) and (5.24′) the estimate (5.37′) follows.

With the basic estimates (5.37)-(5.37′) at our disposal, we proceed as follows. Given any compact set $\mathcal{K}$ in $\boldsymbol{R}_+\setminus e_+(H)$, we choose a subsequence of positive integers $\{j_\nu\}$, $\nu = 1, 2, \ldots$, such that $\sum_\nu \varepsilon_{j_\nu}^2 < \infty$, where $\varepsilon_j = \varepsilon_j(\mathcal{K})$ is the sequence of positive numbers for which the estimate (5.37′) holds. We also choose a number $r > s + n/2$. We let

$$\boldsymbol{R}^n_{\mathcal{K}} = \{\xi : \xi \in \boldsymbol{R}^n,\ |\xi|^2 \in \mathcal{K}\}\,.$$

We shall denote by $\mathcal{E}_{\mathcal{K}}$ the set of points ξ in $\boldsymbol{R}^n_{\mathcal{K}}$ for which

$$(5.40)\qquad \sum_{\nu=1}^{\infty} \int_{\boldsymbol{R}^n} |\phi_{j_\nu}(x, \xi) - \phi_{j_{\nu+1}}(x, \xi)|^2 (1 + |x|)^{-2r}\, dx = \infty\,.$$

Thus, if $\xi \in \boldsymbol{R}^n_{\mathcal{K}}\setminus \mathcal{E}_{\mathcal{K}}$, it follows from (5.40) that

$$\exists \lim_{\nu\to\infty} \phi_{j_\nu}(\cdot, \xi) \qquad \text{in } L^{2,-r}(\boldsymbol{R}^n_x)\,.$$

We claim that the set $\mathcal{E}_{\mathcal{K}}$ has a Lebesgue measure zero, and that, moreover, for any $k^2 \in \mathcal{K}$ the set: $\Lambda_k = \{\omega : \omega \in \Sigma,\ k\omega \in \mathcal{E}_{\mathcal{K}}\}$ is a null set with respect to the measure $d\omega$. Indeed, let $k^2 \in \mathcal{K}$. Using the monotone convergence theorem and the estimate (5.37′), we have

$$\int_{\Sigma}\left(\lim_{p\to\infty}\sum_{\nu=1}^{p}\int_{\boldsymbol{R}^n}|\phi_{j_\nu}(x, k\omega)-\phi_{j_{\nu+1}}(x, k\omega)|^2(1+|x|)^{-2r}\,dx\right)d\omega =$$

$$= \lim_{p\to\infty}\sum_{\nu=1}^{p}\int_{\boldsymbol{R}^n}\left(\int_{\Sigma}|\phi_{j_\nu}(x, k\omega)-\phi_{j_{\nu+1}}(x, k\omega)|^2\,d\omega\right)(1+|x|)^{-2r}\,dx \leqslant$$

$$\leqslant\left(\sum_{\nu=1}^{\infty}\varepsilon_{j_\nu}^2\right)\int_{\boldsymbol{R}^n}(1+|x|)^{-2r+2s}\,dx < \infty\,.$$

This implies that for any fixed k,

$$\sum_{j=1}^{\infty}\int_{\boldsymbol{R}^n}|\phi_{j_\nu}(x, k\omega)-\phi_{j_{\nu+1}}(x, k\omega)|^2(1+|x|)^{-2r}\,dx < \infty$$

for almost all ω in Σ, which proves that Λ_k is a null set with respect to $d\omega$. Similarly, using the monotone convergence theorem and the estimate (5.37′), we find that

$$\text{(5.42)}\qquad \int_{\boldsymbol{R}^n_{\mathcal{K}}}\left(\sum_{\nu=1}^{\infty}\int_{\boldsymbol{R}^n}|\phi_{j_\nu}(x, \xi)-\phi_{j_{\nu+1}}(x, \xi)|^2(1+|x|)^{-2r}\,dx\right)d\xi =$$

$$= \sum_{\nu=1}^{\infty}\int_{\boldsymbol{R}^n}\left(\int_{\boldsymbol{R}^n_{\mathcal{K}}}|\phi_{j_\nu}(x, \xi)-\phi_{j_{\nu+1}}(x, \xi)|^2\,d\xi\right)(1+|x|)^{-2r}\,dx \leqslant$$

$$\leqslant\left(\sum_{\nu=1}^{\infty}\varepsilon_{j_\nu}^2\right)\left(\frac{1}{2}\int_{\mathcal{K}}\lambda^{n/2-1}\,d\lambda\right)\int_{\boldsymbol{R}^n}(1+|x|)^{-2r+2s}\,dx < \infty\,.$$

From (5.42) it follows that $\mathcal{E}_{\mathcal{K}}$ is a null set with respect to the Lebesgue measure on $\boldsymbol{R}^n$.

Let now $\mathcal{K}_i$, $i = 1, 2, \ldots$, be an increasing sequence of compact sets in $\boldsymbol{R}_+\diagdown e_+(H)$ such that $\bigcup_i \mathcal{K}_i = \boldsymbol{R}_+\diagdown e_+(H)$. Let r be a number $> n/2 + s$. By the result just established for $\mathcal{K} = \mathcal{K}_1$, it follows that there exists a subsequence $\{\phi_{j_\nu^1}\}$ of the sequence $\{\phi_j(x, \xi)\}$ such that $\exists \lim\limits_{\nu\to\infty}\phi_{j_\nu^1}(\cdot\,, \xi)$ in $L(^{2,-r}\boldsymbol{R}^n_x)$ for every ξ in $\boldsymbol{R}^n_{\mathcal{K}_1}\diagdown\mathcal{E}_{\mathcal{K}_1}$ where $\mathcal{E}_{\mathcal{K}_1}$ is a null set in $\boldsymbol{R}^n$ whose intersection with any sphere $|\xi| = k$ is a set of measure zero with respect to the Lebesgue measure on the sphere. Applying the same result to the sequence $\{\phi_{j_\nu^1}\}$ and set $\mathcal{K}_2$, it follows that there exists a subsequence $\{\phi_{j_\nu^2}\}$ of $\{\phi_{j_\nu^1}\}$ such that $\exists \lim_{\nu\to\infty} \phi_{j_\nu^2}(\cdot, \xi)$ in $L^{2,-r}(\boldsymbol{R}^n_x)$ for every ξ in $\boldsymbol{R}^n_{\mathcal{K}_2}\setminus\mathcal{E}_{\mathcal{K}_2}$, $\mathcal{E}_{\mathcal{K}_2}$ a null set of the same type as $\mathcal{E}_{\mathcal{K}_1}$ Continuing in this manner, extracting successively sequences of functions $\{\phi_{j_\nu^i}\}\supset\{\phi_{j_\nu^{i+1}}\}$ (i-th subsequence $\{\phi_{j_\nu^i}(\cdot, \xi)\}_{\nu=1}^{\infty}$ converges in $L^{2,-r}(\boldsymbol{R}^n_x)$ for every $\xi\in\boldsymbol{R}^n_{\mathcal{K}_i}\setminus\mathcal{E}_{\mathcal{K}_i}$), and applying a diagonal process, we obtain the following.

LEMMA 5.3. *Let $\{\phi_j(x, \xi)\}$, $j = 1, 2, \ldots$, be the sequence of functions defined by (5.20). Then, given any number $r > n/2 + s$* [16], *there is a subsequence of functions $\{\phi_{j_\nu}\}$ such that*

$$\exists \lim_{\nu\to\infty} \phi_{j_\nu}(\cdot, \xi) \qquad \text{in } L^{2,-r}(\boldsymbol{R}^n_x) \text{ [17]} \tag{5.43}$$

for every $\xi\in\boldsymbol{R}^n\setminus\mathcal{N}$, $\xi\notin\mathcal{E}$, where $\mathcal{E}$ is a null set in $R^n\setminus\mathcal{N}$ having the property that its intersection with any sphere $|\xi| = k$ is a null set with respect to the Lebesgue measure on the sphere.

We have seen already that $\phi_j(x, \xi)$ is a Hölder continuous function of x and that $\phi_j(\cdot, \xi)\in\mathcal{H}_2^{\mathrm{loc}}(\boldsymbol{R}^n_x)$ for every $\xi\in\boldsymbol{R}^n\setminus\mathcal{N}_j$. More precisely, since $\phi_j(x, \xi)$ is a solution of the differential equation (5.21), and since $|V_j|\leqslant W$ for $j = 1, 2, \ldots$, with W satisfying (5.15), it follows upon application of Lemma 5.1 (compare (5.22)) that the following inequality holds:

$$|||\phi_j(\cdot, \xi)|||_{\theta,-r} + \|\phi_j(\cdot, \xi)\|_{2,-r}\leqslant\gamma\|\phi_j(\cdot, \xi)\|_{0,-r} \tag{5.44}$$

for any $\xi\in\boldsymbol{R}^n\setminus\mathcal{N}_j$, $j = 1, 2, \ldots$, where γ is a constant depending only on ξ, θ, r and W.

With the aid of Lemma 5.3 we shall complete the proof as follows. We fix a number $r > n/2 + s$, and use the lemma to extract a subsequence $\{\phi_{j_\nu}(x, \xi)\}$ verifying (5.43). We then apply (5.44) to the functions $\phi_{j_\nu}(\cdot, \xi)$ which are well defined for any fixed $\xi\in\boldsymbol{R}^n\setminus(\mathcal{N}\cup\mathcal{E})$, $\mathcal{E}$ the null set in the lemma, for $\forall\nu\geqslant\nu_0(\xi)$. Since, by (5.43), the sequence $\{\phi_{j_\nu}(\cdot, \xi)\}_{\nu\geqslant\nu_0}$ is bounded in $L^{2,-r}(\boldsymbol{R}^n)$, we have

$$|||\phi_{j_\nu}(\cdot, \xi)|||_{\theta,-r} + \|\phi_{j_\nu}(\cdot, \xi)\|_{2,-r}\leqslant C_\xi \tag{5.45}$$

for $\forall\nu\geqslant\nu_0(\xi)$, where C_ξ is some constant depending on ξ but not on ν. The estimate (5.45) implies in particular that for any fixed $\xi\in\boldsymbol{R}^n\setminus(\mathcal{N}\cup\mathcal{E})$

[16] We can actually assume that $r > (n + 1)/2$, since $s > \frac{1}{2}$ can be chosen as close to $\frac{1}{2}$ as we please.

[17] We recall that if ξ belongs some compact set $K\subset R^n\setminus\mathcal{N}$, $\mathcal{N}$ defined by (5.16)′, then $\varphi_j(x, \xi)$ is a well defined function of x for all $j\geqslant j_0(K)$.

the sequence $\{\phi_{j_\nu}(x,\xi)\}_{\nu\geqslant\nu_0(\xi)}$ is an equicontinuous sequence of functions on any compact subset of $\boldsymbol{R}^n_x$. This and the convergence in norm (5.43) imply that the sequence $\{\phi_{j_\nu}(x,\xi)\}_{\nu\geqslant\nu_0(\xi)}$ converges pointwisely for every x, the convergence being uniform on compact subsets of $\boldsymbol{R}^n_x$.

We now define

$$(5.46)\qquad \begin{cases} \phi_-(x,\xi)=\lim\limits_{\nu\to\infty}\phi_{j_\nu}(x,\xi) & \text{for } x\in\boldsymbol{R}^n,\ \xi\in\boldsymbol{R}^n\diagdown(\mathcal{N}\cup\mathcal{E}),\\ \phi_-(x,\xi)\equiv 0 & \text{for } x\in\boldsymbol{R}^n,\ \xi\in\mathcal{E}.\end{cases}$$

We shall show that $\phi_-(x,\xi)$ has all the properties of the family of generalized eigenfunctions ϕ_- claimed in Theorem 5.1. To this end observe first that $\phi_-(x,\xi)$ is a Hölder continuous function of x and also that $\phi_-(\cdot,\xi)\in\mathcal{H}_2^{\mathrm{loc}}(\boldsymbol{R}^n_x)$ for every fixed $\xi\in\boldsymbol{R}^n\diagdown(\mathcal{N}\cup\mathcal{E})$. More precisely, since by (5.45) the sequence $\phi_{j_\nu}(\cdot,\xi\}$, $\nu\geqslant\nu_0(|\xi|)$, is bounded in $C^{\theta,-r}(\boldsymbol{R}^n_x)$ as well as in $\mathcal{H}_{2,-r}(\boldsymbol{R}^n_x)$, it follows from (5.46) that $\phi_-(\cdot,\xi)\in C^{\theta,-r}(\boldsymbol{R}^n_x)\cap\mathcal{H}_{2,-r}(\boldsymbol{R}^n_x)$ for every $\xi\notin\mathcal{N}\cup\mathcal{E}$. Moreover, the boundedness of $\{\phi_{j_\nu}\}$ in $\mathcal{H}_{2,-r}(\boldsymbol{R}^n_x)$ and the pointwise convergence (5.46) imply that

$$(5.47)\qquad \phi_-(\cdot,\xi)=w-\lim_{\nu\to\infty}\phi_{j_\nu}(\cdot,\xi)\qquad\text{in } \mathcal{H}_{2,-r}(\boldsymbol{R}^n_x)\,.$$

Since $\phi_{j_\nu}(x,\xi)$ satisfies the differential equation (5.21), it follows upon passage to the limit: $\nu\to\infty$, using (5.47) and (5.26′), that

$$(5.48)\qquad \big(-\Delta_x+V(x)-|\xi|^2\big)\phi_-(x,\xi)=0\qquad\text{in } \boldsymbol{R}^n_x$$

for every $\xi\in\boldsymbol{R}^n\diagdown(\mathcal{N}\cup\mathcal{E})$. Since $\phi_-(x,\xi)\equiv 0$ for $\xi\in\mathcal{E}$, we have established that for every fixed $\xi\in\boldsymbol{R}^n\diagdown\mathcal{N}$ the function $\phi_-(x,\xi)$ is in $C^{\theta,-r}\cap\mathcal{H}_{2,-r}$ and satisfies the differential equation (5.48). This shows that ϕ_- has property (ii) of the theorem.

Next, introduce polar coordinates in the ξ-space, $\xi=k\omega$, and write

$$(5.49)\qquad \begin{aligned} \psi_j(x,k,\omega)&=\phi_j(x,k\omega)\,,\qquad j=1,2,\ldots,\\ \psi_-(x,k,\omega)&=\phi_-(x,k\omega)\,.\end{aligned}$$

Since $\phi_j(x,\xi)$ is a well defined continuous function on $\boldsymbol{R}^n\times(\boldsymbol{R}^n\diagdown\mathcal{N}_j)$, it follows that for any compact set $\mathfrak{I}\subset\boldsymbol{R}_+\diagdown e_+(H)^{\frac12}$ the functions $\psi_j(x,k,\omega)$ are well defined continuous functions on $\boldsymbol{R}^n\times\mathfrak{I}\times\Sigma$ for $\forall\ \geqslant j_0$. Consider $\psi_j(x,k,\omega)$ as a vector valued function $\psi_j(x,k,\cdot)$, defined for $(x,k)\in\boldsymbol{R}^n\times\big(\boldsymbol{R}_+\diagdown e_+(H_j)^{\frac12}\big)$ and taking values in $L^2(\Sigma)$. The estimate (5.37′) shows that

$$(5.50)\qquad \exists s-\lim_{j\to\infty}\psi_j(x,k,\cdot)\qquad\text{in } L^2(\Sigma)\,,$$

uniformly in (x, k) on compact subsets of $\boldsymbol{R}^n \times (\boldsymbol{R}_+ \setminus e_+(H)^{\frac{1}{2}})$. Now, from (5.46) and (5.49) it follows (since $\Lambda_k = \{\omega : \omega \in \Sigma,\ k\omega \in \mathcal{E}\}$ is a null set with respect to $d\omega$) that for every fixed x and k

$$\lim_{\nu \to \infty} \psi_{j_\nu}(x, k, \omega) = \psi_-(x, k, \omega) \tag{5.50$'$}$$

for almost all ω in Σ. From (5.50) and (5.50′) it follows that for every (x, k) in $\boldsymbol{R}^n \times (\boldsymbol{R}_+ \setminus e_+(H)^{\frac{1}{2}})$ the function $\psi_-(x, k, \omega)$ belongs to $L^2(\Sigma)$, and that the vector valued function $\psi_-(x, k, \cdot)$, with values in $L^2(\Sigma)$, is a continuous function of x and k. This establishes property (iii) of the theorem.

Since $\phi_-(x, \xi)$ is a measurable function in (x, ξ) it follows readily from the validity of (iii) that $\phi_-(x, \xi) \in L^2_{\text{loc}}(\boldsymbol{R}^n \times (\boldsymbol{R}^n \setminus \mathcal{N}))$. Thus, property (i) of the theorem holds.

For any $g \in L^2(\Sigma)$, define

$$\phi^g_-(x, k) = \int_\Sigma \psi_-(x, k, \omega) g(\omega)\, d\omega\,. \tag{5.51}$$

From the continuity of the vector valued function $\psi(x, k, \cdot)$ it follows that $\phi^g_-(x, k)$ is a well defined continuous function of x and k on $\boldsymbol{R}^n \times (\boldsymbol{R}^n \setminus e_+(H)^{\frac{1}{2}})$. Reintroduce the functions $\phi^g_j(x, k)$ defined by (5.38), or by

$$\phi^g_j(x, k) = \int_\Sigma \psi_j(x, k, \omega) g(\omega)\, d\omega\,, \qquad j = 1, 2, \ldots, \tag{5.51$'$}$$

$\big(\phi^g_j(x, k)$ is a continuous function on $\boldsymbol{R}^n \times (\boldsymbol{R}_+ \setminus e_+(H_j)^{\frac{1}{2}})\big)$. From (5.50), (5.50′), (5.51) and (5.51′) it follows that

$$\lim_{j \to \infty} \phi^g_j(x, k) = \phi^g_-(x, k)\,, \tag{5.52}$$

uniformly in (x, k) on compact subsets of $\boldsymbol{R}^n \times (\boldsymbol{R}_+ \setminus e_+(H)^{\frac{1}{2}})$. Recall that (for a fixed k) the function $\phi^g_j(x, k)$ is in $C^{\theta, -s}(\boldsymbol{R}^n_x) \cap \mathcal{H}_{2,-s}(\boldsymbol{R}^n_x)$. Moreover, by (5.24), the sequence $\{\phi^g_j(\cdot, k)\}_{j \geq j_0(k)}$ (for $k \in \boldsymbol{R}_+ \setminus e_+(H)^{\frac{1}{2}}$) is bounded in $C^{\theta, -s}(\boldsymbol{R}^n_x)$ as well as in $\mathcal{H}_{2,-s}(\boldsymbol{R}^n_x)$. This and (5.52) imply that $\phi^g_-(x, k) \in C^{\theta, -s}(\boldsymbol{R}^n_x) \cap \mathcal{H}_{2,-s}(\boldsymbol{R}^n_x)$.

Finally, using the representation (5.28) for ϕ^g_j and the limit relations (5.27) and (5.52), we find that, for any $k \in \boldsymbol{R}_+ \setminus e_+(H)^{\frac{1}{2}}$,

$$\begin{aligned}
\phi^g_-(\cdot, k) &= \lim_{j \to \infty} \phi^g_j(\cdot\ , k) && \text{in } \mathcal{H}_{2,-s}(\boldsymbol{R}^n_x) \\
&= \lim_{j \to \infty} [\phi^g_0(\cdot, k) - \tilde{T}_j(k^2)\phi^g_0(\cdot, k)] && \text{in } \mathcal{H}_{2,-s}(\boldsymbol{R}^n_x) \\
&= \phi^g_0(\cdot, k) - R^+(k^2)\, V \phi^g_0(\cdot, k)\,.
\end{aligned}$$

This establishes the representation (5.18), proving also that $\phi^{g}_{-}(x, \xi)$ is a solution of the differential equation (5.19). Hence, we have shown that property (iv) of the theorem also holds. The proof of Theorem 5.1 is now complete.

For future reference we note the following result which follows from the proof of Theorem 5.1.

THEOREM 5.2. *Let* $H = -\Delta + V$ *be a Schrödinger operator with potential* V *verifying* (5.15). *Let* $\{\phi_{\pm}(x, \xi)\}$ *be the two families of generalized eigenfunctions associated with* H *by Theorem* 5.1 (*thus in particular* $\phi_{\pm}(x, \xi) \in L^2_{\text{loc}}\big(\boldsymbol{R}^n_x \times \times (\boldsymbol{R}^n_\xi \setminus \mathcal{N}(H))\big)$. *Consider an approximating sequence of Schrödinger operators* $H_j = -\Delta + V_j$, $j = 1, 2, \ldots$, *where the sequence of potentials* $\{V_j\}$ *verifies the following conditions.* (i) $V_j \in L^{2,s_0}(\boldsymbol{R}^n)$ *for some* $s_0 > \frac{1}{2}$. (ii) $V_j(x) \to V(x)$ *for almost all* x. (iii) $|V_j(x)| \leqslant W(x)$ *for* $j = 1, 2, \ldots$, *where* $W(x)$ *is a function satisfying condition* (5.15). *Let* $\{\phi_{j\pm}(x, \xi)\}$ *be the two families of generalized eigenfunctions associated with* H_j. *Then,* $\phi_{j\pm}(x, \xi)$ *are continuous functions of* x *and* ξ *on* $\boldsymbol{R}^n \times (\boldsymbol{R}^n \setminus \mathcal{N}(H_j))$ *such that* $\phi_{j\pm} \to \phi_{\pm}$ *in* $L^2_{\text{loc}}\big(\boldsymbol{R}^n \times (\boldsymbol{R}^n \setminus \mathcal{N}(H))\big)$ *as* $j \to \infty$. *That is, if* $\mathcal{K}$ *any compact set in* $\boldsymbol{R}_+ \setminus e_+(H)$ *and* $r \in \boldsymbol{R}_+$, *then*

$$\lim_{j \to \infty} \int\limits_{|x| \leqslant r} \int\limits_{|\xi|^2 \in \mathcal{K}} |\phi_{\pm}(x, \xi) - \phi_{j\pm}(x, \xi)|^2 \, dx \, d\xi = 0 \, . \tag{5.53}$$

PROOF. it follows from a previous observation (see Remark 2 above) that the families of generalized eigenfunctions $\phi_{\pm}(x, \xi)$ of H are uniquely defined by the conditions of Theorem 5.1 $\Big($as functions in $L^2_{\text{loc}}\big(\boldsymbol{R}^n \times (\boldsymbol{R}^n \setminus \mathcal{N}(H))\big)\Big)$. To verify (5.53) we apply the proof of Theorem 5.1 to our approximating sequence $\{H_j\}$. We observe that in the proof of Theorem 5.1 the generalized eigenfunction $\phi_{j-}(x, \xi)$ was denoted by $\phi_j(x, \xi)$. We have shown there that $\phi_j(x, \xi)$ is a continuous function on $\boldsymbol{R}^n \times (\boldsymbol{R}^n \setminus \mathcal{N}(H_j))$, and that the sequence $\{\phi_j\}$ satisfies the estimates (5.37′). These estimates imply that

$$\lim_{j, m \to \infty} \int\limits_{|x| \leqslant r} \int\limits_{|\xi|^2 \in \mathcal{K}} |\phi_{m-}(x, \xi) - \phi_{j-}(x, \xi)|^2 \, dx \, d\xi = 0 \, . \tag{5.54}$$

Since, by (5.46), $\phi_{j_\nu}(x, \xi) \to \phi_-(x, \xi)$ for almost all (x, ξ) in $\boldsymbol{R}^n \times (\boldsymbol{R}^n \setminus \mathcal{N}(H))$ (for some subsequence $\{j_\nu\}$), it follows from (5.54) that (5.53) holds for ϕ_-. The proof of (5.53) for ϕ_+ is similar.

6. – The eigenfunction expansion theorem.

Before discussing the eigenfunction expansion theorem we establish a simpler spectral property which holds for Schrödinger operators having a potential of class SR. We show that the spectrum of any such operator H is absolutely continuous on any interval not containing eigenvalues of H. More precisely, we have

THEOREM 6.1. *Let $L^2(\boldsymbol{R}^n)_{ac}$ be the absolute continuity subspace* ([18]) *of $L^2(\boldsymbol{R}^n)$ with respect to $H = -\Delta + V$ where V is a potential of class SR. Let $L^2(\boldsymbol{R}^n)_p$ be the closed subspace in $L^2(\boldsymbol{R}^n)$ spanned by the eigenfunctions of H. Then,*

$$L^2(\boldsymbol{R}^n) = L^2(\boldsymbol{R}^n)_{ac} \oplus L^2(\boldsymbol{R}^n)_p . \tag{6.1}$$

REMARK. – Theorem 6.1 under different conditions on the potential was proved by many authors (see for instance [8], [10] and references given there). In the following we denote by $\{E_\lambda\}$ the spectral resolution of the identity associated with H, and we denote by $E(\mathcal{B})$ the projection associated with a Borel set $\mathcal{B} \subset \boldsymbol{R}$. We also denote by P_{ac} the projection of $L^2(\boldsymbol{R}^n)$ onto $L^2(\boldsymbol{R}^n)_{ac}$. With this notation Theorem 6.1 asserts that

$$P_{ac} = E(\boldsymbol{R}_+ \setminus e_+(H)) . \tag{6.1$'$}$$

PROOF OF THEOREM 6.1. We have to show that a necessary and sufficient condition for $(E_\lambda f, f)$ to be absolutely continuous on $\boldsymbol{R}$ is for f to be orthogonal to $L^2(\boldsymbol{R}^n)_p$. The necessity part of the condition is obvious. To prove sufficiency we observe that if $f \perp L^2(\boldsymbol{R}^n)_p$ then $(E_\lambda f, f)$ is a continuous function on $\boldsymbol{R}$, vanishing for $\lambda \leqslant 0$. Because of this and because $e_+(H)$ is a discrete set, to prove sufficiency it is enough to show that the function $(E_\lambda f, f)$ is absolutely continuous on any compact interval in $\boldsymbol{R}_+ \setminus e_+(H)$.

Thus, consider an interval $[a, b] \subset \boldsymbol{R}_+ \setminus e_+(H)$. We shall use the well known formula

$$([E_b - E_a]f, f) = \lim_{\varepsilon \to +0} \frac{1}{2\pi i} \int_a^b ([R(\lambda + i\varepsilon) - R(\lambda - i\varepsilon)]f, f)\, d\lambda , \tag{6.2}$$

which holds for any $f \in L^2(\boldsymbol{R}^n)$ (see [3], p. 1202). Assume that $f \in L^{2,s}(\boldsymbol{R}^n)$ for some $s > \frac{1}{2}$. It follows from (6.2) and Theorem 4.2 that

$$([E_b - E_a]f, f) = \frac{1}{2\pi i} \int_a^b ([R^+(\lambda) - R^-(\lambda)]f, f)\, d\lambda . \tag{6.2$'$}$$

Hence, it follows from (6.2′) that $(E_\lambda f, f)$ is continuously differentiable on

([18]) See section 1 for definition.

$\boldsymbol{R}_+ \setminus e_+(H)$ and that

$$\frac{d}{d\lambda}(E_\lambda f, f) = \frac{1}{2\pi i}\left([R^+(\lambda) - R^-(\lambda)]f, f\right) \tag{6.2''}$$

for any $f \in L^{2,s}(\boldsymbol{R}^n)$, $s > \frac{1}{2}$. Since the set of functions f for which $(E_\lambda f, f)$ is an absolutely continuous function of λ on $[a, b]$ is closed in $L^2(\boldsymbol{R}^n)$ (see [7]), it follows from (6.2″) that $(E_\lambda f, f)$ is absolutely continuous on every compact interval in $\boldsymbol{R}_+ \setminus e_+(H)$ for *all* $f \in L^2(\boldsymbol{R}^n)$. This establishes the theorem.

We now turn to the eigenfunction expansion theorem.

THEOREM 6.2. *Let $H = -\Delta + V$ be a Schrödinger operator with potential V verifying condition* (5.15). *Let $\phi_\pm(x, \xi)$ be the family of generalized eigenfunctions introduced in Theorem* 5.1. *There exist two bounded linear maps*

$$\mathcal{F}_\pm : L^2(\boldsymbol{R}^n) \to L^2(\boldsymbol{R}^n),$$

with the following properties:

(i) $\ker(F_\pm) = L^2(\boldsymbol{R}^n)_p$. *The restriction of $\mathcal{F}_\pm$ to $L^2(\boldsymbol{R}^n)_{ac}$ is a unitary operator from $L^2(\boldsymbol{R}^n)_{ac}$ onto $L^2(\boldsymbol{R}^n)$.*

(ii) *For any $f \in L^2(\boldsymbol{R}^n)$*

$$(\mathcal{F}_\pm f)(\xi) = (2\pi)^{-n/2} \lim_{N\to\infty} \int_{|x|<N} f(x)\overline{\phi_\pm(x, \xi)}\, dx \qquad \text{in } L^2(\boldsymbol{R}^n_\xi), \tag{6.3}$$

and

$$(\mathcal{F}^*_\pm f)(x) = (2\pi)^{-n/2} \lim_{j\to\infty} \int_{K_j} f(\xi)\phi_\pm(x, \xi)\, d\xi \qquad \text{in } L^2(\boldsymbol{R}^n_x), \tag{6.3'}$$

where K_j is an increasing sequence of compacts such that $\bigcup_j K_j = \boldsymbol{R}^n \setminus \mathcal{N}$ ($\mathcal{N}$ the null set defined by (5.16)′).

(iii) *Let P_{ac} be the projection of $L^2(\boldsymbol{R}^n)$ onto $L^2(\boldsymbol{R}^n)_{ac}$, then*

$$(P_{ac}H)f = (\mathcal{F}^*_\pm M_{|\cdot|^2}\mathcal{F}_\pm)f \qquad \text{for } \forall f \in \mathcal{D}(H) \tag{6.4}$$

where $M_{|\cdot|^2}$ denotes the multiplication operator by $|\xi|^2$.

We note that (6.4) and (1.2) yield the following

COROLLARY. *Let H_{ac} be the restriction of H to the reducing subspace $L^2(\boldsymbol{R}^n)_{ac}$. Then,*

$$H_{ac} = U_\pm H_0 U^*_\pm$$

where $U_{\pm}$ *are unitary operators from* $L^2(\boldsymbol{R}^n)$ *onto* $L^2(\boldsymbol{R}^n)_{\mathrm{ac}}$ *given by* $U_{\pm} = \mathcal{F}_{\pm}^{*}\mathcal{F}$ *(*$\mathcal{F}$ *denotes the Fourier map).*

PROOF OF THEOREM 6.2. We denote by $L_0^2(\boldsymbol{R}^n)$ the class of L^2 functions with compact support in $\boldsymbol{R}^n$. For $f \in L_0^2(\boldsymbol{R}^n)$, we set

$$(\mathcal{F}_{\pm}f)(\xi) = (2\pi)^{-n/2}\int_{\boldsymbol{R}^n} f(x)\,\overline{\phi_{\pm}(x,\xi)}\,dx\,. \tag{6.5}$$

Using the properties of the generalized eigenfunctions $\phi_{\pm}(x,\xi)$ (see Theorem 5.1) it follows that $(\mathcal{F}_{\pm}f)(\xi)$ is well defined for any $\xi \in \boldsymbol{R}^n\setminus\mathcal{N}$ and that $\mathcal{F}_{\pm}f \in L^2_{\mathrm{loc}}(\boldsymbol{R}^n\setminus\mathcal{N})$. Let $\{E_\lambda\}$ be the spectral resolution of the identity of H. We shall show that

$$([E_b - E_a]f, f) = \int_{a<|\xi|^2<b} |(\mathcal{F}_{\pm}f)(\xi)|^2\,d\xi \tag{6.6}$$

for any interval $[a, b] \subset \boldsymbol{R}_+\setminus e_+(H)$.

We shall first establish (6.6) under the additional assumption that V is a bounded function with compact support ([19]). To this end we introduce the resolvent operator $R(z) = (H-z)^{-1}$ and compute the Fourier transform of $R(z)f$ $\big(f \in L_0^2(\boldsymbol{R}^n),\ \mathrm{Im}\, z \neq 0\big)$. We claim that

$$\mathcal{F}\big(R(z)f\big)(\xi) = \big(|\xi|^2 - z\big)^{-1}\tilde{f}(\xi, z) \tag{6.7}$$

where

$$\tilde{f}(\xi, z) = (2\pi)^{-2/n}\int_{\boldsymbol{R}^n} f(x)\,\overline{\big(\phi_0(x,\xi) - R(\bar{z})[V(\cdot)\phi_0(\cdot,\xi)](x)\big)}\,dx\,, \tag{6.7'}$$

$\phi_0(x,\xi) = \exp(ix\cdot\xi)$. Indeed, suppose first that f is of the form $f = (H-z)u$, with $u \in C_0^\infty(\boldsymbol{R}^n)$, then

$$\begin{aligned}
(2\pi)^{n/2}\mathcal{F}\big(R(z)f\big)(\xi) &= (2\pi)^{n/2}\hat{u}(\xi) = \\
&= \big(|\xi|^2 - z\big)^{-1}\int_{\boldsymbol{R}^n}(-\Delta - z)\,u(x)\cdot\exp[-ix\cdot\xi]\,dx \\
&= \big(|\xi|^2 - z\big)^{-1}\int_{\boldsymbol{R}^n}[(H-z)\,u(x) - V(x)\,u(x)]\,\overline{\phi_0(x,\xi)}\,dx \\
&= \big(|\xi|^2 - z\big)^{-1}\Big[\hat{f}(\xi) - \int_{\boldsymbol{R}^n} V(x)\big(R(z)f\big)(x)\cdot\overline{\phi_0(x,\xi)}\,dx\Big] \\
&= \big(|\xi|^2 - z\big)^{-1}\Big[\hat{f}(\xi) - \int_{\boldsymbol{R}^n} f(x)\,\overline{R(\bar{z})[V(\cdot)\phi_0(\cdot\,.\,\xi)](x)}\,dx\Big]\,.
\end{aligned}$$

([19]) In this special case (6.6) is proved in [5], [23] and [2]. The proof which we give here (mainly for the sake of completeness) is different in some details.

This yields (6.7) for f in the class of functions

$$\mathfrak{L} = \{f : f = (H - z)u,\ u \in C_0^\infty(\boldsymbol{R}^n)\}\,.$$

Since $\mathfrak{L}$ is dense in $L_0^2(\boldsymbol{R}^n)$ (in the L^2 norm) it follows that (6.7) holds for $\forall f \in L_0^2(\boldsymbol{R}^n)$.

Next, we fix an interval $[a, b]$ in $\boldsymbol{R}_+ \setminus e_+(H)$ and set

$$Q_\pm[a, b] = \{z : z \in \boldsymbol{C}, \quad a \leqslant \operatorname{Re} z \leqslant b, \quad 0 < \pm \operatorname{Im} z \leqslant 1\}\,.$$

For any $f \in L_0^2(\boldsymbol{R}^n)$ we consider the function $\tilde{f}(\xi, z)$ defined by (6.7′). It is readily seen that $\tilde{f}(\xi, z)$ is a continuous function of ξ and z on $\boldsymbol{R}^n \times Q_\pm[a, b]$. Moreover, using Theorem 4.2 it follows that

$$\exists \lim_{\varepsilon \to +0} \tilde{f}(\xi, \lambda \pm i\varepsilon) = \tilde{f}_\pm(\xi, \lambda) \tag{6.8}$$

uniformly for $a \leqslant \lambda \leqslant b$. Since V has a compact support it follows from the proof of Theorem 5.1 that the generalized eigenfunctions $\phi_\pm(x, \xi)$ are continuous functions given by (5.2). Hence, combining (6.5), (6.7′), (6.8) and (5.2), it follows that

$$\tilde{f}_\pm(\xi, |\xi|^2) = (\mathcal{F}_\pm f)(\xi) \qquad \text{for } a \leqslant |\xi|^2 \leqslant b\,. \tag{6.9}$$

We shall also need the estimate

$$\|\tilde{f}(\cdot, z)\|_{L^2(\boldsymbol{R}^n)} \leqslant C \qquad \text{for } \forall z \in Q_\pm[a, b] \tag{6.10}$$

where C is some constant. To prove (6.10) we multiply both sides of (6.7′) by a function $g(\xi) \in C_0^\infty(\boldsymbol{R}^n)$ and integrate over $\boldsymbol{R}^n$. We get

$$\int_{\boldsymbol{R}^n} \tilde{f}(\xi, z)\, g(\xi)\, d\xi = \int_{\boldsymbol{R}^n} f(x)\, \hat{g}(x)\, dx - \\ - \int_{\boldsymbol{R}^n} f(x) \cdot R(z)[V(\cdot)\hat{g}(\cdot)](x)\, dx\,. \tag{6.11}$$

From (6.11) it follows that

$$\left|\int_{\boldsymbol{R}^n} \tilde{f}(\xi, z)\, g(\xi)\, d\xi\right| \leqslant \|f\|_0 \|\hat{g}\|_0 + \|f\|_{0,s} \|R(z)[V\hat{g}]\|_{0,-s} \tag{6.12}$$

for any $s \geqslant 0$. Choosing $s > \frac{1}{2}$ and applying Theorem 4.2, using the fact that V is a bounded function with compact support, we have

$$\|R(z)[V\hat{g}]\|_{0,-s} \leqslant C_1 \|V\hat{g}\|_{0,s} \leqslant C_2 \|g\|_0 \tag{6.12'}$$

for $\forall z \in Q_\pm[a, b]$ where C_1 and C_2 are constants which do not depend on z. Combining (6.12) and (6.12′) it follows that there exists a constant C such that

$$\left| \int_{\mathbf{R}^n} \tilde{f}(\xi, z) g(\xi)\, d\xi \right| \leqslant C \|g\|_0$$

for $\forall g \in C_0^\infty(\mathbf{R}^n)$ and $\forall z \in Q_\pm[a, b]$. This yields (6.10).

We are in a position to prove (6.6) (special case). We shall use once more formula (6.2) which we rewrite in the following form:

$$([E_b - E_a]f, f) = \lim_{\varepsilon \to +0} \frac{\varepsilon}{\pi} \int_a^b \|R(\lambda \pm i\varepsilon) f\|^2 \, d\lambda \,. \tag{6.13}$$

Assuming as above that $f \in L_0^2(\mathbf{R}^n)$ and noting that by Parseval's formula

$$\int_a^b \|R(\lambda \pm i\varepsilon) f\|^2 \, d\lambda = \int_a^b \|\mathcal{F}[R(\lambda \pm i\varepsilon) f\|^2 \, d\lambda \,, \tag{6.14}$$

it follows from (6.13), (6.14) and (6.7) that

$$([E_b - E_a]f, f) = \lim_{\varepsilon \to +0} \int_{\mathbf{R}^n} G_{\pm\varepsilon}(\xi)\, d\xi \,, \tag{6.15}$$

where we set

$$G_{\pm\varepsilon}(\xi) = \frac{\varepsilon}{\pi} \int_a^b [(|\xi|^2 - \lambda)^2 + \varepsilon^2]^{-1} |\tilde{f}(\xi, \lambda \pm i\varepsilon)|^2 \, d\lambda \,. \tag{6.15'}$$

Observing that $\tilde{f}(\xi, z)$ is a continuous function of ξ and z on $\mathbf{R}^n \times Q_\pm[a, b]$ which admits a continuous extension to $\mathbf{R}^n \times \overline{Q_\pm[a, b]}$, it follows from (6.8) and well known properties of the Poisson integral that

$$\lim_{\varepsilon \to +0} G_{\pm\varepsilon}(\xi) = \begin{cases} f_\pm(\xi, |\xi|^2) & \text{if } a < |\xi|^2 < b \,, \\ 0 & \text{if } |\xi|^2 < a \text{ or } |\xi|^2 > b \,. \end{cases} \tag{6.16}$$

Now, to evaluate the limit (6.15) we break the integral on $\mathbf{R}^n$ into two parts: an integral on the sphere $|\xi| < A$ where we choose $A > \sqrt{b}$, and an integral on $|\xi| \geqslant A$. Since the family of functions $\{G_{\pm\varepsilon}(\xi)\}_{0<\varepsilon\leqslant 1}$ is uniformly bounded on $|\xi| < A$, it follows from (6.15′) and (6.16) that

$$\lim_{\varepsilon\to+0} \int_{|\xi|<A} G_{\pm\varepsilon}(\xi)\, d\xi = \int_{a<|\xi|^2<b} \tilde{f}_{\pm}(\xi, |\xi|^2)\, d\xi \,. \tag{6.16′}$$

It follows from (6.15′) that, for $|\xi| \geqslant A$,

$$|G_{\pm\varepsilon}(\xi)| \leqslant \frac{\varepsilon}{\pi(A^2 - b)^2} \int_a^b |\tilde{f}(\xi, \lambda \pm i\varepsilon)|^2\, d\lambda \,.$$

Hence, using the bound (6.10) we have

$$\left| \int_{|\xi|\geqslant A} G_{\pm\varepsilon}(\xi)\, d\xi \right| \leqslant \frac{\varepsilon}{\pi(A^2 - b)^2} \int_a^b \int_{|\xi|\geqslant A} |\tilde{f}(\xi, \lambda \pm i\varepsilon)|^2\, d\xi\, d\lambda \leqslant \tag{6.16″}$$

$$\leqslant \varepsilon \frac{(b-a)\, C^2}{\pi(A^2 - b)^2} \to 0 \qquad \text{as } \varepsilon \to 0 \,.$$

Combining (6.15), (6.16′), (6.16″) and (6.9), we get

$$([E_b - E_a] f, f) = \int_{a<|\xi|^2<b} |\tilde{f}_{\pm}(\xi, |\xi|^2)|^2\, d\xi = \int_{a<|\xi|^2<b} |(\mathcal{F}_{\pm} f)(\xi)|^2\, d\xi \,.$$

This proves (6.6) under the additional assumption that V is a bounded function with compact support.

To establish (6.6) in the general case (i.e. assuming only that V verifies (5.15)) we approximate V by a sequence of functions V_j with compact supports, $j = 1, 2, \ldots$, defined as follows

$$V_j(x) = \begin{cases} V(x) & \text{if } |x| \leqslant j \text{ and } |V(x)| \leqslant j \,, \\ 0 & \text{otherwise} \,. \end{cases}$$

We denote by H_j the Schrödinger operator $-\Delta + V_j$. We set $R_j(z) = (H_j - z)^{-1}$ and denote by $R_j^{\pm}(\lambda)$ the boundary values of $R_j(z)$ on $\boldsymbol{R}_+$ (defined by Theorem 4.2). We denote by $\{E_\lambda^j\}$ the spectral resolution of the identity of H_j.

Let $\{\phi_{j\pm}(x, \xi)\}$ be the generalized eigenfunctions associated with H_j. For $f \in L_0^2(\boldsymbol{R}^n)$ define

$$(\mathcal{F}_{\pm}^j f)(\xi) = (2\pi)^{-n/2} \int_{\boldsymbol{R}^n} f(x)\, \overline{\phi_{j\pm}(x, \xi)}\, dx \,. \tag{6.17}$$

Since $\phi_{j\pm}(x,\xi)$ is a continuous function of x and ξ on $\boldsymbol{R}^n\times(\boldsymbol{R}^n\setminus\mathcal{N}(H_j))$ (see Theorem 5.2), it follows that $(\mathcal{F}^j_\pm f)(\xi)$ is continuous on $\boldsymbol{R}^n\setminus\mathcal{N}(H_j)=$ $=\{\xi\colon \xi\in\boldsymbol{R}^n\setminus\{0\},\ |\xi|^2\notin e_+(H_j)\}$. As before let $[a,b]$ be an interval contained in $\boldsymbol{R}_+\setminus e_+(H)$. Applying Theorem 4.3 to the sequence $\{H_j\}$ it follows that there exists an integer j_0 such that $[a,b]\cap e_+(H_j)=\emptyset$ for $\forall j\geq j_0$ [20]. Since V_j is a bounded function with compact support it follows from the result established above that (6.6) holds for H_j, i.e., we have

$$([E^j_b-E^j_a]f,f)=\int\limits_{a<|\xi|^2<b}|(\mathcal{F}^j_\pm f)(\xi)|^2\,d\xi \tag{6.18}$$

for $\forall f\in L^2_0(\boldsymbol{R}^n)$ and $\forall j\geq j_0$. Now, applying Theorem 4.3 and using (6.2)$'$ we find that

$$\lim_{j\to\infty}([E^j_b-E^j_a]f,f)=\lim_{j\to\infty}\frac{1}{2\pi i}\int_a^b([R^+_j(\lambda)-R^-_j(\lambda)]f,f)\,d\lambda= \tag{6.19}$$

$$=\frac{1}{2\pi i}\int_a^b([R^+(\lambda)-R^-(\lambda)]f,f)\,d\lambda=([E_b-E_a]f,f)\,.$$

On the other hand, using Theorem 5.2 it follows from (6.5), (6.17) and (5.53) that

$$\lim_{j\to\infty}\int\limits_{a<|\xi|^2<b}|(\mathcal{F}^j_\pm f)(\xi)|^2\,d\xi=\int\limits_{a<|\xi|^2<b}|(\mathcal{F}_\pm f)(\xi)|^2\,d\xi \tag{6.19$'$}$$

for any $f\in L^2_0(\boldsymbol{R}^n)$. Combining (6.18), (6.19) and (6.19$'$), the relation (6.6) follows.

From (6.6) it follows readily that if $\mathcal{O}$ is any open set in $\boldsymbol{R}_+\setminus e_+(H)$, then

$$(E(\mathcal{O})f,f)=\int\limits_{|\xi|^2\in\mathcal{O}}|(\mathcal{F}_\pm f)(\xi)|^2\,d\xi\,. \tag{6.6$'$}$$

Applying (6.6$'$) with $\mathcal{O}=\boldsymbol{R}_+\setminus e_+(H)$ and using Theorem 6.2, we get in particular that

$$\|P_{\mathrm{ac}}f\|=\|\mathcal{F}_\pm f\| \tag{6.20}$$

for $\forall f\in L^2_0(\boldsymbol{R}^n)$ $(P_{\mathrm{ac}}$ the projection of $L^2(\boldsymbol{R}^n)$ onto $L^2(\boldsymbol{R}^n)_{\mathrm{ac}})$.

[20] It actually follows from Kato [6] that $e_+(H_j)=\emptyset$ for $\forall j$.

From (6.20) it follows that the mapping $f \to \mathcal{F}_\pm f$ is a contraction from $L_0^2(\boldsymbol{R}^n)$ into $L^2(\boldsymbol{R}^n)$. Extending the map by continuity to the whole of $L^2(\boldsymbol{R}^n)$, we denote also the extended map by $\mathcal{F}_\pm$. Thus $\mathcal{F}_\pm$ is a contraction from $L^2(\boldsymbol{R}^n)$ into $L^2(\boldsymbol{R}^n)$ which (by (6.20)) satisfies the relation

$$\mathcal{F}_\pm P_{\mathrm{ac}} = \mathcal{F}_\pm . \tag{6.21}$$

We shall show that the linear map $\mathcal{F}_\pm$ which we have just defined has all the properties described in the theorem.

First we observe that (6.21) and (6.1) imply that $\ker(\mathcal{F}_\pm) = L^2(\boldsymbol{R}^n)_p$ and that $\mathcal{F}_\pm$ is an isometry from $L^2(\boldsymbol{R}^n)_{\mathrm{ac}}$ into $L^2(\boldsymbol{R}^n)$. This yields part (i) of the theorem except for the fact that $\mathrm{range}(\mathcal{F}_\pm) = L^2(\boldsymbol{R}^n)$. This last property we shall establish later on.

To prove (ii) it would suffice to establish the relation (6.3′). Now, let f and g be two functions in $L_0^2(\boldsymbol{R}^n)$ with $\operatorname{supp} f \subset \boldsymbol{R}^n \setminus \mathcal{N}(H)$. We have

$$\begin{aligned}(\mathcal{F}_\pm^* f, g) = (f, \mathcal{F}_\pm g) &= (2\pi)^{-n/2}\int_{\boldsymbol{R}^n} f(\xi)\Big(\int_{\boldsymbol{R}^n}\overline{g(x)\phi_\pm(x,\xi)}\,dx\Big)\,d\xi = \\ &= (2\pi)^{-n/2}\int_{\boldsymbol{R}^n}\overline{g(x)}\Big(\int_{\boldsymbol{R}^n} f(\xi)\phi_\pm(x,\xi)\,d\xi\Big)\,dx ,\end{aligned} \tag{6.22}$$

where the change of order of integrations is justified by Fubini's theorem. Hence, it follows from (6.22) that

$$(\mathcal{F}_\pm^* f)(x) = (2\pi)^{-n/2}\int_{\boldsymbol{R}^n} f(\xi)\,\phi_\pm(x,\xi)\,d\xi \tag{6.22′}$$

for any $f \in L_0^2(\boldsymbol{R}^n)$ with $\operatorname{supp} f \subset \boldsymbol{R}^n \setminus \mathcal{N}(H)$. Clearly, (6.3′) follows from (6.22′) by continuity. This yields (ii).

Let $g \in L^2(\boldsymbol{R}^n)_{\mathrm{ac}}$. Since $(E_\lambda g, g)$ is an absolutely continuous function of λ, it follows from (6.6) that

$$(E_\lambda g, g) = \int_{|\xi|^2<\lambda} |(\mathcal{F}_\pm g)(\xi)|^2\,d\xi , \qquad \lambda > 0 . \tag{6.23}$$

Hence, if $g \in L^2(\boldsymbol{R}^n)_{\mathrm{ac}} \cap \mathfrak{D}(H)$ it follows, using (6.23), that

$$\begin{aligned}(Hg, g) &= \int_0^\infty \lambda\, d(E_\lambda g, g) = \int_{\boldsymbol{R}^n} |\xi|^2 |(\mathcal{F}_\pm g)(\xi)|^2\,d\xi \\ &= (M_{|\cdot|^2}\mathcal{F}_\pm g, \mathcal{F}_\pm g) = (\mathcal{F}_\pm^* M_{|\cdot|^2}\mathcal{F}_\pm g, g) .\end{aligned} \tag{6.24}$$

Since $L^2(\boldsymbol{R}^n)_{\mathrm{ac}} \cap \mathfrak{D}(H)$ is dense in $L^2(\boldsymbol{R}^n)_{\mathrm{ac}}$, it follows from (6.24) that

$$Hg = \mathcal{F}_{\pm}^{*} M_{|\cdot|^2} \mathcal{F}_{\pm} g \,. \tag{6.25}$$

Applying (6.25) with $g = P_{\mathrm{ac}} f$, $f \in \mathfrak{D}(H)$, and using (6.21) we find that

$$HP_{\mathrm{ac}} f = \mathcal{F}_{\pm}^{*} M_{|\cdot|^2} \mathcal{F}_{\pm} f \tag{6.26}$$

for any $f \in \mathfrak{D}(H)$. This yields (6.4), proving (iii).

To complete the proof of the theorem we still have to show that $\operatorname{range}(\mathcal{F}_{\pm}) = L^2(\boldsymbol{R}^n)$. For the proof of this result let us note the following relation

$$\mathcal{F}_{\pm}(E(\mathfrak{J}) f)(\xi) = \chi_{\mathfrak{J}}(|\xi|^2)(\mathcal{F}_{\pm} f)(\xi) \,, \tag{6.27}$$

which holds for any compact interval $\mathfrak{J} \subset \boldsymbol{R}_+ \setminus e_+(H)$ and any $f \in L^2(\boldsymbol{R}^n)$ ($\chi_{\mathfrak{J}}(\lambda)$ denotes the characteristic function of the interval $\mathfrak{J}$ on $\boldsymbol{R}$). Indeed (6.27) follows readily from the relation (6.26) just proved, observing also that (by a computation similar to (6.24)) we have

$$\|Hg\|^2 = \|M_{|\cdot|^2} \mathcal{F}_{\pm} g\|^2 \qquad \text{for } \forall g \in L^2_{\mathrm{ac}}(\boldsymbol{R}^n) \cap \mathfrak{D}(H) \,.$$

We shall prove that $\operatorname{range}(\mathcal{F}_{\pm}) = L^2(\boldsymbol{R}^n)$ by showing that $\ker(\mathcal{F}_{\pm}^{*}) = \{0\}$. I.e., let

$$\mathcal{F}_{\pm}^{*} f = 0 \qquad \text{for some } f \in L^2(\boldsymbol{R}^n) \,. \tag{6.28}$$

We shall show that $f = 0$. To this end we observe that it follows from (6.28) and (6.27) (by taking adjoints) that

$$\mathcal{F}_{\pm}^{*}(\chi_{\mathfrak{J}}(|\cdot|^2) f) = E(\mathfrak{J})\, \mathcal{F}_{\pm}^{*} f = 0 \,, \tag{6.28'}$$

for any interval $\mathfrak{J} = [a, b]$ in $\boldsymbol{R}_+ \setminus e_+(H)$. Hence, using (6.22′) it follows from (6.28′) that

$$\int\limits_{a<|\xi|^2<b} f(\xi)\, \phi_{\pm}(x, \xi)\, d\xi = 0 \qquad \text{for } \forall x \in \boldsymbol{R}^n \,. \tag{6.29}$$

Next, we introduce polar coordinates: $k = |\xi| \in \boldsymbol{R}_+$, $\omega = \xi/|\xi| \in \Sigma$, and set $\psi_{\pm}(x, k, \omega) = \phi_{\pm}(x, k\omega)$. We define

$$u_{\pm}(x, k) = \int\limits_{\Sigma} f(k\omega)\, \psi_{\pm}(x, k, \omega)\, d\omega \,. \tag{6.30}$$

Since the vector valued function $\psi_{\pm}(x, k, \cdot)$, with values in $L^2(\Sigma)$, is continuous on $\boldsymbol{R}^n \times (\boldsymbol{R}_+ \setminus e_+(H)^{\frac{1}{2}})$ (see Theorem 5.1), it follows readily from (6.30) that for each fixed x the function $u_{\pm}(x, k)$ is defined for almost all k in $\boldsymbol{R}_+$ and that $u_{\pm}(x, \cdot) \in L^1_{\text{loc}}(\boldsymbol{R}_+ \setminus e_+(H)^{\frac{1}{2}})$. Moreover, from (6.29), (6.30) and Fubini's theorem, it follows that

$$\int_{\sqrt{a}}^{\sqrt{b}} k^{n-1} u_{\pm}(x, k)\, dk = 0 \tag{6.31}$$

for any $x \in \boldsymbol{R}^n$ where $[a, b]$ is any interval in $\boldsymbol{R}_+ \setminus e_+(H)$. From (6.31) and the arbitrariness of a and b it follows that, for each fixed x,

$$u_{\pm}(x, k) = 0 \qquad \text{for almost all } k \text{ in } \boldsymbol{R}_+ . \tag{6.32}$$

More precisely, we claim that there exists a null set Λ in $\boldsymbol{R}_+$ such that

$$u_{\pm}(x, k) = 0 \qquad \text{on } \boldsymbol{R}^n \times (\boldsymbol{R}_+ \setminus \Lambda) . \tag{6.32'}$$

Indeed, let $\{x_j\}$ be a dense sequence of points in $\boldsymbol{R}^n$. It follows readily from (6.32) that there exists a null set Λ in $\boldsymbol{R}_+$ such that $u_{\pm}(x_j, k) = 0$ for $\forall k \in \boldsymbol{R}_+ \setminus \Lambda$ and $\forall j$ (we choose Λ so that $u_{\pm}(x, k)$ is well defined for $x \in \boldsymbol{R}^n$ for each $k \in \boldsymbol{R}_+ \setminus \Lambda$). Since $u_{\pm}(x, k)$ is a continuous function of x for each $k \in \boldsymbol{R}_+ \setminus \Lambda$, it follows by continuity that (6.32′) holds.

We consider now the function

$$u(x, k) = \int_{\Sigma} f(k\omega) \exp[ik\omega \cdot x]\, d\omega . \tag{6.33}$$

It was assumed implicitly above that Λ is chosen so that for each fixed $k \in \boldsymbol{R}_+ \setminus \Lambda$ the function $f(k\omega)$ is a well defined function of class $L^2(\Sigma)$. Hence, for each such k, $u(x, k)$ is a well defined function of class $\mathcal{H}_{2,-s}(\boldsymbol{R}^n) \cap C^\infty(\boldsymbol{R}^n)$ for any $s > \frac{1}{2}$ (by Lemma 5.2). Next observe that u and $u_{\pm}$ are connected by the relation

$$u_{\pm}(x, k) = u(x, k) - R^{\mp}(k^2)[V(\cdot)u(\cdot, k)](x) . \tag{6.34}$$

Indeed, this follows from Theorem 5.1 since (6.34) is nothing else but the relation (5.18) written for $g(\omega) = f(k\omega)$ (with the notation of Theorem 5.1 we have $u_{\pm} = \phi^g_{\pm}$, $u = \phi^g_0$). Hence, combining (6.34) with (6.32′) we find that

$$u(x, k) = R^{\mp}(k^2)[V(\cdot)u(\cdot, k)](x) \tag{6.35}$$

for every $k \in \boldsymbol{R}_+ \setminus \Lambda$. Finally, combining (6.35) with the resolvent equation (4.18), we find that

$$(6.36) \qquad u(x, k) = R_0^{\mp}(k^2)[V(\cdot)u(\cdot, k)](x) - R_0^{\mp}(k^2)[V(\cdot)R^{\mp}(k^2)V(\cdot)u(\cdot, k)](x) \\ = R_0^{\mp}(k^2)[V(\cdot)(u(\cdot, k) - R^{\mp}(k^2)V(\cdot)u(\cdot, k))](x) = 0 ,$$

for $k \in \boldsymbol{R}_+ \setminus \Lambda$.

From (6.36) and (6.33) it follows (by integration) that

$$\int\limits_{c<|\xi|<d} f(\xi)\exp[i\xi\cdot x]\,d\xi = 0$$

for any $0 < c < d$, which implies (by Fourier transform) that $f(\xi) = 0$ for almost all ξ. This establishes that $\operatorname{range}(\mathcal{F}_\pm) = L^2(\boldsymbol{R}^n)$ and completes the proof of the theorem.

7. – The eigenfunction expansion theorem and scattering theory.

In this section we shall show that the eigenfunction expansion theorem furnishes a useful tool for the study of certain problems in scattering theory. We shall consider here briefly the two problems mentioned in the introduction: the problem of existence and completeness of wave operators $W_\pm$ (associated with a pair of Schrödinger operators $H = -\Delta + V$ and $H_0 = -\Delta$), and the problem of existence of the scattering matrix $S(k)$.

As was mentioned in the Introduction, the first problem was studied by many authors. Among the various solutions given to the problem the following solution was given recently in [4].

THEOREM 7.1. *Let $H = -\Delta + V$ be a Schrödinger operator with potential V verifying condition* (5.15). *Then the wave operators*:

$$W_\pm = s - \lim_{t\to\pm\infty} \exp[itH]\exp[-itH_0] ,$$

exist and are complete.

We shall indicate here a proof of Theorem 7.1 which is based on the eigenfunction expansion theorem. For reasons of brevity we shall not give here a self-contained proof of the theorem, but rather make use of the fact that the theorem is well known in case V has a compact support.

PROOF OF THEOREM 7.1. The proof of existence of wave operators is elementary. It is given in Kuroda [11]. In this connection it should be ob served that condition (5.15) (or even the weaker condition that V is of class SR) implies that

$$V \in L^{2,\beta}(\boldsymbol{R}^n) \qquad \text{for some } \beta > \left(\frac{n}{2} - 1\right),$$

which is the condition in [11] ensuring the existence of wave operators. We also remark that the proof of Kuroda's result follows easily from the formula:

$$\exp[itH]\exp[-itH_0]f = f + i\int_0^t \exp[isH]V\exp[-isH_0]f\,ds\,,$$

which holds for $f \in \mathfrak{D}(H_0)$, and from the observation that

$$\int_{-\infty}^{\infty} \|V\exp[-itH_0]f\|dt < \infty\,, \tag{7.1}$$

for f lying in some subset of $\mathfrak{D}(H_0)$, dense in $L^2(\boldsymbol{R}^n)$.

Next, let $\mathcal{F}_\pm$ be the generalized Fourier maps associated with H, defined in Theorem 6.2. We shall establish the formula

$$W_\pm = \mathcal{F}_\pm^* \mathcal{F}\,, \tag{7.2}$$

$\mathcal{F}$ the ordinary Fourier map. Since by Theorem 6.2 the operators $\mathcal{F}_\pm^*\mathcal{F}$ are isometries from $L^2(\boldsymbol{R}^n)$ onto $L^2(\boldsymbol{R}^n)_{\text{ac}}$, the relation (7.2) will imply in particular that the $W_\pm$ are complete.

To prove (7.2) we approximate V by a sequence of potentials $\{V_j\}$, $j = 1, 2, \ldots$, defined by

$$V_j(x) = \begin{cases} V(x) & \text{if } |x| \leqslant j \text{ and } |V_j(x)| \leqslant j\,, \\ 0 & \text{otherwise}\,. \end{cases} \tag{7.3}$$

We consider the Schrödinger operators $H_j = -\Delta + V_j$, and denote by $W_{j\pm}$ the wave operators of the pair (H_j, H_0). We denote by $\mathcal{F}_{j\pm}$ the generalized Fourier maps which correspond to H_j, and set $U_{j\pm} = \mathcal{F}_{j\pm}^*\mathcal{F}$. We shall show that the following relations hold.

(i) $W_{j\pm} = U_{j\pm}$ for $j = 1, 2, \ldots$.

(ii) $W_{j\pm} \to W_\pm$ strongly in $L^2(\boldsymbol{R}^n)$.

(iii) $U_{j\pm} \to \mathcal{F}_\pm^*\mathcal{F}$ weakly in $L^2(\boldsymbol{R}^n)$.

It is clear that these relations imply (7.2).

Now (i), which is the relation (7.2) in case V is a bounded function with a compact support, follows from the results of Ikebe [5]. To be precise, Ikebe considered in [5] only Schrödinger operators acting on functions on $\boldsymbol{R}^3$. However, given Theorem 6.2, the extension of Ikebe's results to $\boldsymbol{R}^n$ is immediate. We shall not repeat Ikebe's argument here (see also [23], [2] and [22]).

The proof of (ii) is elementary and is based on the estimate (7.1). We refer to [2], p. 301, for a complete proof.

We shall prove (iii). Let $f \in L_0^2(\boldsymbol{R}^n)$, and let g be a function in $L^2(\boldsymbol{R}^n)$ such that $\mathcal{F}g$ has a compact support in $\boldsymbol{R}^n \setminus \mathcal{N}$ ($\mathcal{N}$ the null set (5.16′)). Recalling the definition of $U_{j\pm}$, we have

$$(7.4) \quad |(f, [\mathcal{F}_\pm^* \mathcal{F} - U_{j\pm}]g)| = |([\mathcal{F}_\pm - \mathcal{F}_{j\pm}]f, \mathcal{F}g)| \leqslant \|\chi[\mathcal{F}_\pm - \mathcal{F}_{j\pm}]f\| \, \|\mathcal{F}g\| ,$$

where $\chi(x)$ is the characteristic function of $\operatorname{supp} \mathcal{F}g$. From Theorem 5.2 it follows (using (5.53) together with the definition of $\mathcal{F}_\pm f$ and $\mathcal{F}_{j\pm} f$) that

$$(7.5) \qquad \lim_{j \to \infty} \|\chi[\mathcal{F}_\pm f - \mathcal{F}_{j\pm} f]\| = 0 .$$

Hence, combining (7.4) and (7.5) we get

$$(7.6) \qquad \lim_{j \to \infty} (f, [\mathcal{F}_\pm^* \mathcal{F} - U_{j\pm}]g) = 0 .$$

Since (7.6) holds for all pairs (f, g) such that f belongs to a certain dense set in $L^2(\boldsymbol{R}^n)$, and g belongs to another such set, it follows by continuity (since $U_{j\pm}$ are isometries) that (7.6) holds for all f and g in $L^2(\boldsymbol{R}^n)$. This yields (iii) and completes the proof.

Next we establish the existence of the scattering matrix for the class of Schrödinger operators with potentials verifying (5.15). In this generality the result we prove seems to be new.

THEOREM 7.2. *Let H be a Schrödinger operator with potential V verifying* (5.15). *Let $S = W_+^* W_-$ be the scattering operator of (H, H_0) ($W_\pm$ the wave operators introduced in Theorem* 7.1). *Set $\hat{S} = \mathcal{F} S \mathcal{F}^*$. Then there exists an operator valued function $\mathrm{S}(k)$, defined for $k \in \boldsymbol{R}_+ \setminus e_+(H)^{\frac{1}{2}}$ and taking values in the class of unitary operators on $L^2(\Sigma)$ (Σ the sphere $|\omega| = 1$ in $\boldsymbol{R}^n$), such that the following relation holds:*

$$(7.7) \qquad (\hat{S}f)(k, \omega) = \mathrm{S}(k) f(k, \cdot)(\omega)$$

for $f(k, \omega) \in L^2(\boldsymbol{R}_\xi^n)$, where the equality (7.7) *holds in $L^2(\Sigma)$ for almost all k. (Here k, ω are polar coordinates of ξ).*

The operator valued function $\mathcal{S}(k)$, *called the scattering matrix, has the following properties.*

(i) $k \to \mathcal{S}(k)$ *is a continuous map from* $\boldsymbol{R}_+ \setminus e_+(H)^{\frac{1}{2}}$ *into* $B(L^2(\Sigma), L^2(\Sigma))$.

(ii) *Let* $\mathcal{S}_-(k) = I - \mathcal{S}(k)$ *where* I *is the identity in* $B(L^2(\Sigma), L^2(\Sigma))$.

Then $\mathcal{S}_-(k)$ *is a compact operator for any* $k \in \boldsymbol{R}_+ \setminus e_+(H)^{\frac{1}{2}}$. *Moreover,* $\mathcal{S}_-(k)$ *has a distribution kernel* $s_-(k, \omega, \omega')$ *given symbolically by*

$$s_-(k, \omega, \omega') \sim \frac{i}{2k}\left(\frac{k}{2\pi}\right)^{n-1}\int_{\boldsymbol{R}^n}\phi_-(x, k\omega)\,V(x)\exp[-ik\omega'\cdot x]\,dx \tag{7.8}$$

where $\phi_-(x, \xi)$ *is the generalized eigenfunction of* H *defined in Theorem* 5.1. *Here the symbolic representation* (7.8) *has the following meaning. For any two functions* $g(\omega)$ *and* $h(\omega)$ *in* $L^2(\Sigma)$, *and for every* $k \in \boldsymbol{R}_+ \setminus e_+(H)^{\frac{1}{2}}$,

$$(\mathcal{S}_-(k)g, h) = \frac{i}{2k}\left(\frac{k}{2\pi}\right)^{n-1}\int_{\boldsymbol{R}^n}\phi_-^g(x, k)\,V(x)\overline{\phi_0^h(x, k)}\,dx\,, \tag{7.9}$$

where ϕ_-^g *and* ϕ_0^h *are defined by* (5.17).

PROOF. For a Schrödinger operator with a potential V which decays rapidly at infinity the theorem is well known. In particular the theorem is known to hold if V is a function with compact support satisfying (5.15) (see for instance [2]). In this case it follows also that $\mathcal{S}_-(k)$ is an integral operator

$$(\mathcal{S}_-(k)f)(\omega') = \int_{\Sigma} s_-(k, \omega, \omega')f(\omega)\,d\omega\,, \qquad f \in L^2(\Sigma)\,,$$

with a continuous kernel $s_-(k, \omega, \omega')$ given by the right hand side of (7.8). (To be precise, when V has a compact support it follows from the results of [2] that $\mathcal{S}_-(k)$ is a well defined integral operator only for $k \in \boldsymbol{R}_+ \setminus F$ where F is some unspecified closed set of measure zero in $\boldsymbol{R}_+$. However, using the explicit form of the kernel, it follows readily from our results that the kernel is well defined by (7.8) for all $k \in \boldsymbol{R}_+ \setminus e_+(H)^{\frac{1}{2}}$. From this it follows that $\mathcal{S}_-(k)$ admits an extension as a continuous operator valued function to the whole of $\boldsymbol{R}_+ \setminus e_+(H)^{\frac{1}{2}}$.)

In order to prove the theorem in the general case we shall (as in the proof of Theorem 7.1) approximate V by a sequence of potentials $\{V_j\}$ with compact supports, defined by (7.3). We shall consider the sequence of Schrödinger operators $H_j = -\Delta + V_j$; denoting by $W_{j\pm}$ the wave operators of the pair (H_j, H_0).

We set $S_j = W_{j+}^* W_{j-}$. Our first observation is that

$$s-\lim_{j\to\infty} S_j = S . \tag{7.10}$$

Indeed, from the proof of Theorem 7.1 (or rather from [2], p. 301) we have

$$W_{j\pm} \to W_\pm \quad \text{strongly} . \tag{7.11}$$

From (7.11) it follows, taking adjoints, that

$$W_{j\pm}^* \to W_\pm^* \quad \text{weakly} . \tag{7.11'}$$

Hence, combining (7.11) and (7.11′) we find that

$$w-\lim_{j\to\infty} S_j = w-\lim_{j\to\infty} W_{j+}^* W_{j-} = W_+^* W_- = S . \tag{7.10'}$$

Since S_j and S are isometries, it follows from (7.10′) that (7.10) holds.

We shall now use the fact that Theorem 7.2 is known to hold for H_j (see previous remarks). We denote by $\mathcal{S}_j(k)$ the scattering matrix associated with H_j and set $\mathcal{S}_{j-}(k) = I - \mathcal{S}_j(k)$. By the theorem applied to H_j it follows that $\mathcal{S}_{j-}(k)$ is a continuous function on $\boldsymbol{R}_+ \setminus e_+(H_j)^{\frac{1}{2}}$ with values in $B(L^2(\Sigma), L^2(\Sigma))$. For a fixed k, $\mathcal{S}_{j-}(k)$ is a compact (actually integral) operator on $L^2(\Sigma)$ such that

$$(\mathcal{S}_{j-}(k)g, h) = \frac{i}{2k}\left(\frac{k}{2\pi}\right)^{n-1} \int_{\boldsymbol{R}^n} \phi_j^g(x, k)\, V_j(x) \overline{\phi_0^h(x, k)}\, dx \tag{7.12}$$

for any two functions $g(\omega)$, $h(\omega)$ in $L^2(\Sigma)$, where

$$\phi_j^g(x, k) = \int_\Sigma \phi_{j-}(x, k\omega) g(\omega)\, d\omega , \qquad \phi_0^h(x, k) = \int_\Sigma \exp[ik\omega\cdot x] h(\omega)\, d\omega .$$

(Here $\phi_{j-}(x, \xi)$ is the generalized eigenfunction which corresponds to H_j by Theorem 5.1). Inserting in (7.12) for ϕ_j^g its expression (5.18) (or (5.23)), we find that the bilinear form associated with $\mathcal{S}_{j-}(k)$ admits also the following representation:

$$\begin{aligned}(\mathcal{S}_{j-}(k)g, h) &= \frac{i}{2k}\left(\frac{k}{2\pi}\right)^{n-1} \int_{\boldsymbol{R}^n} \phi_0^g(x, k)\, V_j(x) \overline{\phi_0^h(x, k)}\, dx - \\ &\quad - \frac{i}{2k}\left(\frac{k}{2\pi}\right)^{n-1} \int_{\boldsymbol{R}^n} R_j^+(k^2)[V_j(\cdot)\phi_0^g(\cdot, k)](x)\cdot V_j(x) \overline{\phi_0^h(x, k)}\, dx ,\end{aligned} \tag{7.12'}$$

where $R_j^+(k^2)$ stands for the boundary value of the resolvent of H_j at k^2 (defined by Theorem 4.2; $R_j^+(k^2) \in B(L^{2,s}, \mathcal{H}_{2,-s})$ for $s > \frac{1}{2}$).

We have

$$(7.13) \qquad \lim_{j\to\infty} (S_{j-}(k)g, h) = \left(\frac{i}{2k}\right)\left(\frac{k}{2\pi}\right)^{n-1}\int_{R^n} \phi_0^g(x, k)\, V(x)\overline{\phi_0^h(x, k)}\, dx -$$

$$-\frac{i}{2k}\left(\frac{k}{2\pi}\right)^{n-1}\int_{R^n} R^+(k^2)[V(\cdot)\phi_0^g(\cdot, k)](x)\cdot V(x)\overline{\phi_0^h(x, k)}\, dx\,,$$

where (7.13) holds uniformly in g, h and k, for g and h in the unit ball of $L^2(\Sigma)$ and $k \in \mathcal{K}$, $\mathcal{K}$ any compact set in $\boldsymbol{R}_+ \setminus e_+(H)^{\frac{1}{2}}$ (observe that, by Theorem 4.3, $R_j^+(k^2)$ is well defined on $\mathcal{K}$ for all j sufficiently large).

To prove (7.13) we recall that the multiplication operators V and V_j are compact operators from $\mathcal{H}_{2,-s_0}(\boldsymbol{R}^n)$ into $L^{2,s_0}(\boldsymbol{R}^n)$ for some $s_0 > \frac{1}{2}$ (since $V(x)$ is a function of class SR). Moreover, by (4.28) we have

$$(7.14) \qquad \lim_{j\to\infty} V_j = V \qquad \text{in } B(\mathcal{H}_{2,-s_0}, L^{2,s_0})\,.$$

We also note that by Lemma 5.2 (see also remark which follows the lemma) the function $\phi_0^g(x, k)$ belongs to $\mathcal{H}_{2,-s_0}(\boldsymbol{R}^n)$. Moreover, the following estimate holds:

$$(7.15) \qquad \|\phi_0^g(\cdot, k)\|_{2,-s_0} \leq C_0 \|g\|_{L^2(\Sigma)}$$

for $\forall g \in L^2(\Sigma)$ and $\forall k \in \mathcal{K}$ where C_0 is some constant. From (7.14) and (7.15) it follows that

$$(7.16) \qquad \lim_{j\to\infty} V_j\phi_0^g(\cdot, k) = V\phi_0^g(\cdot, k) \qquad \text{in } L^{2,s_0}(\boldsymbol{R}^n)\,,$$

uniformly in g and k for $\|g\| \leq 1$ and $k \in \mathcal{K}$. From (7.16) and Theorem 4.3, we get

$$(7.17) \qquad \lim_{j\to\infty} R_j^+(k^2)\, V_j\phi_0^g(\cdot, k) = R^+(k^2)\, V\phi_0^g(\cdot, k) \qquad \text{in } L^{2,-s_0}(\boldsymbol{R}^n)\,,$$

uniformly in g and k for $\|g\| \leq 1$ and $k \in \mathcal{K}$. Combining now (7.12′), (7.16) and (7.17), we obtain (7.13).

The above considerations show in particular that the right hand side of (7.13) is a well defined bounded bilinear form on $L^2(\Sigma)$ for every fixed k. Hence, for every $k \in \boldsymbol{R}_+ \setminus e_+(H)^{\frac{1}{2}}$, there exists a bounded linear operator in

$B(L^2(\Sigma), L^2(\Sigma))$, which we shall denote by $\mathcal{S}_-(k)$, such that

$$(7.18)\qquad (\mathcal{S}_-(k)g, h) = \left(\frac{i}{2k}\right)\left(\frac{k}{2\pi}\right)^{n-1}\int_{\boldsymbol{R}^n}\phi_0^g(x, k)\, V(x)\overline{\phi_0^h(x, k)}\, dx -$$
$$-\left(\frac{i}{2k}\right)\left(\frac{k}{2\pi}\right)^{n-1}\int_{\boldsymbol{R}^n} R^+(k^2)[V(\cdot)\phi_0^g(\cdot, k)](x)\cdot V(x)\overline{\phi_0^h(x, k)}\, dx\,.$$

We define $\mathcal{S}(k) = I - \mathcal{S}_-(k)$, $k \in \boldsymbol{R}_+ \setminus e_+(H)^{\frac{1}{2}}$. We shall show that $\mathcal{S}(k)$ has the properties of the scattering matrix described in the theorem.

Now, it is clear from the definition of $\mathcal{S}(k)$ and (7.13) that

$$(7.19)\qquad \mathcal{S}(k) = \lim_{j\to\infty} \mathcal{S}_j(k) \qquad \text{in } B(L^2(\Sigma), L^2(\Sigma))\,,$$

where (7.19) holds uniformly in k on any compact subset of $\boldsymbol{R}_+ \setminus e_+(H)^{\frac{1}{2}}$. Since $\mathcal{S}_j(k)$ is a scattering matrix, having the properties of Theorem 7.2 with respect to H_j, it follows from (7.19) that, for each k, the operator $\mathcal{S}(k)$ is unitary and that $\mathcal{S}(k) - I$ is compact. It also follows that the map: $k \to \mathcal{S}(k)$, is a continuous map from $\boldsymbol{R}_+ \setminus e_+(H)^{\frac{1}{2}}$ into $B(L^2(\Sigma), L^2(\Sigma))$. Next, applying (7.7) to $\mathcal{S}_j(k)$ we have

$$(7.20)\qquad (\mathcal{F} S_j \mathcal{F}^* f)(k, \omega) = \mathcal{S}_j(k) f(k, \cdot)(\omega)\,, \qquad j = 1, 2, \ldots,$$

for any $f \in L^2(\boldsymbol{R}^n)$. Letting $j \to \infty$, it follows from (7.20), (7.19) and (7.10) that (7.7) holds for $\mathcal{S}(k)$. Finally, combining (7.18) and (5.18) we find that (7.9) holds. This shows that $\mathcal{S}(k)$ has the desired properties and completes the proof of the theorem.

Appendix A.

In this appendix we shall establish the a-priori weighted estimates (4.4) which had a crucial role in our proof of the validity of the limiting absorption principle (given in section 4). As a matter of fact we shall go here beyond the needs of the present study and shall establish analogous weighted estimates for the class of operators $P(D) - z$ where $P(D)$ is a differential operator of principal type and z is a complex parameter. As was already mentioned in the Introduction, the estimates which we shall give here could be used to extend the spectral and scattering theory results of this paper to self-adjoint realizations of higher order elliptic operators, as well as to certain non-elliptic operators of principal type.

Let $P(D) = P(D_1, \dots, D_n)$ be a partial differential operator with constant coefficients of order m, acting on functions on $\boldsymbol{R}^n$. We denote its principal part by $P_m(D)$. The operator P is said to be of principal type if

$$\text{(A.1)} \qquad \operatorname{grad} P_m(\xi) \neq 0 \quad \text{for } \forall \xi \in \boldsymbol{R}^n \setminus \{0\}\,.$$

P is said to be elliptic if

$$\text{(A.1')} \qquad P_m(\xi) \neq 0 \quad \text{for } \forall \xi \in \boldsymbol{R}^n \setminus \{0\}\,.$$

Clearly an elliptic operator is also an operator of principal type.

We shall say that a number $z \in \boldsymbol{C}$ is a *critical value* of P if there exists a $\xi_0 \in \boldsymbol{R}^n$ such that $P(\xi_0) = z$, $\operatorname{grad} P(\xi_0) = 0$.

We shall denote the set of all critical values of P by $\Lambda_C(P)$. If $P(\xi)$ is a homogeneous polynomial of degree $\geqslant 2$ then $\Lambda_C(P)$ consists of the single point $\{0\}$. In general we have the following

THEOREM. *The set of critical values $\Lambda_C(P)$ is a finite set.*

For a proof of this theorem see Milnor [16], p. 16 (Corollary 2.8).

We shall establish the following

THEOREM A.1. *Let $P(D)$ be a differential operator with constant coefficients of order m and of principal type. Set $m' = m$ if P is elliptic, $m' = m - 1$ otherwise. Let $\mathcal{K}$ be a compact set in $\boldsymbol{C} \setminus \Lambda_C(P)$ and let $s > \frac{1}{2}$. The following estimate holds*

$$\text{(A.2)} \qquad \|u\|_{m',-s} \leqslant C \|(P(D) - z)\, u\|_{0,s}$$

for $\forall u \in \mathcal{H}_m(\boldsymbol{R}^n)$ *where C is some constant not depending on z or u.*

REMARK 1. Theorem A.1 when specialized to the operator $P = -\Delta$ is Lemma 4.1.

REMARK 2. One can prove the following sharper form of Theorem A.1. Let $\Lambda_C(P)_\delta = \{z : z \in \boldsymbol{C},\ \operatorname{dist}(z, \Lambda_C(P)) < \delta\}$. Then for any given $s > \frac{1}{2}$ and $\delta > 0$ there exists a constant $C = C_{s,\delta}$ such that

$$\text{(A.2')} \qquad \sum_{|\alpha| \leqslant m'} (|z| + 1)^{(m-1-|\alpha|)/m} \|D^\alpha u\|_{0,-s} \leqslant C \|(P(D) - z)\, u\|_{0,s}$$

for $\forall u \in \mathcal{H}_m(\boldsymbol{R}^n)$ and $\forall z \in \boldsymbol{C} \setminus \Lambda_C(P)_\delta$.

We reduce the proof of Theorem A.1 to the proof of certain lemmas. The first lemma is an elementary inequality.

LEMMA A.1. *Let $u \in \mathcal{H}_1(\boldsymbol{R})$, $\lambda \in \boldsymbol{C}$ and $s > \frac{1}{2}$. The following inequality holds:*

$$\|u\|_{0,-s} \leqslant c_s \left\| \left(\frac{d}{dx} - \lambda\right) u \right\|_{0,s} \tag{A.3}$$

where

$$c_s = 2\int_0^\infty (1 + x^2)^{-s}\, dx . \tag{A.3$'$}$$

PROOF. We set

$$f(x) = \left(\frac{d}{dx} - \lambda\right) u(x) , \qquad u \in \mathcal{H}_1(\boldsymbol{R}) . \tag{A.4}$$

We may assume without loss of generality that $f \in L_1(\boldsymbol{R})$ (otherwise $\|f\|_{0,s} = \infty$ and (A.3) holds trivially), and that $\operatorname{Re} \lambda \leqslant 0$. Solving (A.4) for u we get

$$u(x) = \int_{-\infty}^{x} f(t) \exp[\lambda(x - t)]\, dt . \tag{A.5}$$

From (A.5) it follows that

$$|u(x)|^2 \leqslant \Big(\int_{-\infty}^{x} |f(t)|\, dt\Big)^2 \leqslant \Big(\int_{-\infty}^{\infty} (1 + t^2)^{-s}\, dt\Big) \int_{-\infty}^{\infty} (1 + t^2)^s |f(t)|^2\, dt . \tag{A.6}$$

Multiplying (A.6) by $(1 + x^2)^{-s}$ and integrating on $\boldsymbol{R}$, we obtain

$$\int_{-\infty}^{\infty} |u(x)|^2 (1 + x^2)^{-s}\, dx \leqslant \Big(\int_{-\infty}^{\infty} (1 + t^2)^{-s}\, dt\Big)^2 \int_{-\infty}^{\infty} |f(t)|^2 (1 + t^2)^s\, dt .$$

This yields the lemma.

Given a polynomial $P(\xi) = P(\xi_1, \dots, \xi_n)$, we set: $P^{(k)}(\xi) = (\partial/\partial \xi_k) P(\xi)$, $k = 1, \dots, m$.

LEMMA A.2. *Let $P(D) = P(D_1, \dots, D_n)$ be a partial differential operator of order m. Then for $\forall u \in \mathcal{H}_m(\boldsymbol{R}^n)$ and any given $s > \frac{1}{2}$, the following inequality holds:*

$$\int_{\boldsymbol{R}^n} (1 + x_j^2)^{-s} |P^{(j)}(D)\, u|^2\, dx \leqslant m^2 c_s^2 \int_{\boldsymbol{R}^n} (1 + x_j^2)^s |P(D)\, u|^2\, dx , \tag{A.7}$$

for $j = 1, \dots, m$, where c_s is the constant (A.3$'$).

PROOF. We shall prove (A.7) for $j = 1$. We write $P(\xi) = P(\xi_1, \xi')$, $\xi' = (\xi_2, \dots, \xi_n)$. For a fixed ξ' we denote by $\lambda_j(\xi')$, $j = 1, \dots, \nu(\xi')$, the roots of $P(\xi_1, \xi')$ in ξ_1 taken according to their multiplicity. We have

$$\text{(A.8)} \qquad P^{(1)}(\xi_1, \xi') = \frac{\partial}{\partial \xi_1} P(\xi_1, \xi') = \sum_{j=1}^{\nu(\xi')} Q_j(\xi_1, \xi')$$

where the Q_j are polynomials in ξ_1,

$$\text{(A.8}'\text{)} \qquad Q_j(\xi_1, \xi') = P(\xi_1, \xi')(\xi_1 - \lambda_j(\xi'))^{-1}.$$

Let, now, $u \in C_0^\infty(\boldsymbol{R}^n)$ and write $u(x) = u(x_1, x')$, $x' = (x_2, \dots, x_n)$. We denote by $\tilde{u}(x_1, \xi')$ the Fourier transform of u with respect to the variable x'. Using the relation (A.8) it follows that for any fixed ξ' with $\nu(\xi') > 0$,

$$\text{(A.9)} \qquad P^{(1)}(D_1, \xi')\tilde{u}(x_1, \xi') = \sum_{j=1}^{\nu(\xi')} Q_j(D_1, \xi')\,\tilde{u}(x_1, \xi').$$

We shall apply now Lemma A.1 to the function $v_j(x_1) = Q_j(D_1, \xi')\tilde{u}(x_1, \xi')$, taking $\lambda = i\lambda_j(\xi')$. From (A.3) and (A.8') it follows that

$$\begin{aligned}\text{(A.10)} \qquad &\int_{-\infty}^{\infty}(1 + x_1^2)^{-s}|Q_j(D_1, \xi')\tilde{u}(x_1, \xi')|^2\,dx_1 \leqslant \\ &\leqslant c_s^2\int_{-\infty}^{\infty}(1 + x_1^2)^{s}|(D_1 - \lambda_j(\xi'))Q_j(D_1, \xi')\tilde{u}(x_1, \xi')|^2\,dx_1 \\ &= c_s^2\int_{-\infty}^{\infty}(1 + x_1^2)^{s}|P(D_1, \xi')\tilde{u}(x_1, \xi')|^2\,dx_1.\end{aligned}$$

Combining (A.9) and (A.10), we get

$$\begin{aligned}\text{(A.11)} \qquad &\int_{-\infty}^{\infty}(1 + x_1^2)^{-s}|P^{(1)}(D_1, \xi')\tilde{u}(x_1, \xi')|^2\,dx_1 \leqslant \\ &\leqslant m^2 c_s^2\int_{-\infty}^{\infty}(1 + x_1^2)^{s}|P(D_1, \xi')\tilde{u}(x_1, \xi')|^2\,dx_1\end{aligned}$$

for any ξ' with $\nu(\xi') > 0$. Since (A.11) holds trivially when $\nu(\xi') = 0$, we have (A.11) for all ξ'. Integrating (A.11) with respect to ξ', using Parseval's formula, we find that

$$\text{(A.12)}\quad \int\limits_{\mathbf{R}^n}(1+x_1^2)^{-s}|P^{(1)}(D)\,u(x)|^2\,dx$$

$$=\int\limits_{\mathbf{R}^{n-1}} d\xi'\int\limits_{-\infty}^{\infty}(1+x_1^2)^{-s}|P^{(1)}(D_1,\xi')\,\tilde{u}(x_1,\xi')|^2\,dx_1$$

$$\leqslant m^2c_s^2\int\limits_{\mathbf{R}^{n-1}} d\xi'\int\limits_{-\infty}^{\infty}(1+x_1^2)^{s}|P(D_1,\xi')\,\tilde{u}(x_1,\xi')|^2\,dx_1$$

$$=m^2c_s^2\int\limits_{\mathbf{R}^n}(1+x_1^2)^{s}|P(D)\,u(x)|^2\,dx\,.$$

This proves the lemma for $\forall u\in C_0^\infty(\mathbf{R}^n)$.

To complete the proof of the lemma we observe first that (A.12) holds also for any function u with compact support in $\mathcal{H}_m(\mathbf{R}^n)$. This follows readily from the result just established and from the fact that such u is a limit in $\mathcal{H}_m(\mathbf{R}^n)$ of a sequence of functions $u_k\in C_0^\infty(\mathbf{R}^n)$ such that the u_k have their supports in some fixed ball. Next, we choose a cutoff function $\zeta\in C_0^\infty(\mathbf{R}^n)$ with $\zeta(x)=1$ for $|x|\leqslant 1$ and set $\zeta_k(x)=\zeta(k^{-1}x)$. By the above (A.12) holds for $u_k=u\zeta_k$. Letting $k\to\infty$ it follows readily that (A.12) holds also for u. (In the last step one assumes without loss of generality that $\frac{1}{2}<s\leqslant 1$.) This establishes the lemma.

LEMMA A.3. *Let $P(D)$ be a differential operator of principal type of order m. Set $m'=m$ if P is elliptic, $m'=m-1$ otherwise. Let $\mathcal{K}$ be a compact set in $\mathbf{C}\setminus\Lambda_C(P)$ and let s be a real number. The following estimate holds*

$$\text{(A.13)}\qquad \|u\|_{m',s}\leqslant C_s\Big(\|(P(D)-z)\,u\|_{0,s}+\sum_{j=1}^{n}\|P^{(j)}(D)\,u\|_{0,s}\Big)$$

for $\forall u\in\mathcal{H}_{m,s}(\mathbf{R}^n)$ and $\forall z\in\mathcal{K}$, where C_s is a constant not depending on z or u.

PROOF. Since P is of principal type it follows that there exist positive constants C and R_0 such that

$$\text{(A.14)}\qquad (1+|\xi|^2)^{m-1}\leqslant C\sum_{j=1}^{n}|P^{(j)}(\xi)|^2\qquad \text{for } |\xi|\geqslant R_0\,.$$

Since $\mathcal{K}\cap\Lambda_C(P)=\emptyset$, we also have

$$\text{(A.14}')\qquad \min_{\substack{|\xi|\leqslant R_0\\ z\in\mathcal{K}}}\Big(|P(\xi)-z|^2+\sum_{j=1}^{n}|P^{(j)}(\xi)|^2\Big)=c_0>0\,.$$

Hence, it follows from (A.14) and (A.14′) that with a possibly larger constant C, we have

$$(1+|\xi|^2)^{m-1}\leqslant C\Big(|P(\xi)-z|^2+\sum_{j=1}^{n}|P^{(j)}(\xi)|^2\Big) \tag{A.15}$$

for all $\xi\in\boldsymbol{R}^n$ and $z\in\mathcal{K}$. If P is also elliptic it follows further that with a different constant C we have

$$(1+|\xi|^2)^{m}\leqslant C\Big(|P(\xi)-z|^2+\sum_{j=1}^{n}|P^{(j)}(\xi)|^2\Big) \tag{A.15′}$$

for all $\xi\in\boldsymbol{R}^n$ and $z\in\mathcal{K}$.

Let now $u(x)$ be a function in $\mathcal{H}_m(\boldsymbol{R}^n)$ with Fourier transform $\hat{u}(\xi)$. Multiplying both sides of (A.15) ((A.15′) if P is elliptic) by $|\hat{u}(\xi)|^2$ and integrating on $\boldsymbol{R}^n$, using Parseval's formula, we find that

$$\sum_{|\alpha|\leqslant m'}\|D^\alpha u\|^2\leqslant\gamma\int_{\boldsymbol{R}^n}(1+|\xi|^2)^{m'}|\hat{u}(\xi)|^2\,d\xi\leqslant\gamma C\Big(\|(P(D)-z)u\|^2+\sum_{j=1}^{n}\|P^{(j)}(D)u\|^2\Big)$$

for $\forall z\in\mathcal{K}$ where γ is a constant depending only on n and m (here and in the following $\|\cdot\|$ denotes the norm in $L^2(\boldsymbol{R}^n)$). This proves (A.13) for $s=0$. To prove (A.13) in the general case fix a real s and introduce the family of weight functions $\varrho_\varepsilon(x)$ on $\boldsymbol{R}^n$, defined for every $\varepsilon>0$ by $\varrho_\varepsilon(x)=(1+|\varepsilon x|^2)^{s/2}$. Observe that

$$|D^\alpha\varrho_\varepsilon(x)|\leqslant C_\alpha\varepsilon^{|\alpha|}\varrho_\varepsilon(x) \tag{A.16}$$

for $x\in\boldsymbol{R}^n$, $\varepsilon>0$ and $|\alpha|\geqslant0$ where C_α are constants given by

$$C_\alpha=\sup_x\big(|D^\alpha\varrho_1(x)|\varrho_1(x)^{-1}\big)\,.$$

Observe also that if $Q(D)$ is a differential operator with constant coefficients of order k, then

$$\|Q(D)(\varrho_\varepsilon u)-\varrho_\varepsilon Q(D)u\|\leqslant\varepsilon K_Q\sum_{|\alpha|\leqslant k-1}\|\varrho_\varepsilon D^\alpha u\| \tag{A.17}$$

for $\forall u\in\mathcal{H}_{m,s}(\boldsymbol{R}^n)$ and $\forall\varepsilon$ such that $0<\varepsilon\leqslant1$ (K_Q is some constant not depending on ε). Indeed, we have

$$Q(D)(\varrho_\varepsilon u)-\varrho_\varepsilon Q(D)u=\sum_{1\leqslant|\alpha|\leqslant k}D^\alpha\varrho_\varepsilon\cdot Q_\alpha(D)u\,, \tag{A.17′}$$

where the Q_α are certain differential operators of orders $\leqslant k-1$. Combining (A.17′) and (A.16) the estimate (A.17) follows.

Let now $u \in \mathcal{H}_{m,s}(\boldsymbol{R}^n)$. Applying the estimate (A.13) for $s=0$ proved above to the function $\varrho_\varepsilon u$, we have

$$\text{(A.18)} \qquad \sum_{|\alpha|\leqslant m'} \|D^\alpha(\varrho_\varepsilon u)\| \leqslant C\Big(\|(P(D)-z)(\varrho_\varepsilon u)\| + \sum_{j=1}^{n} \|P^{(j)}(D)(\varrho_\varepsilon u)\|\Big)$$

for $\forall z \in \mathcal{K}$ (C some constant depending only on P and $\mathcal{K}$). Using (A.17) for $Q(D)=D^\alpha$, $Q(D)=P(D)-z$ and $Q(D)=P^{(j)}(D)$, it follows from (A.18) that for any $u \in \mathcal{H}_{m,s}(\boldsymbol{R}^n)$, $z \in \mathcal{K}$ and $0<\varepsilon\leqslant 1$,

$$\text{(A.19)} \qquad \begin{aligned} \sum_{|\alpha|\leqslant m'} \|\varrho_\varepsilon D^\alpha u\| &\leqslant \sum_{|\alpha|\leqslant m'} \|D^\alpha(\varrho_\varepsilon u)\| + \varepsilon K_1 \sum_{|\alpha|\leqslant m'-1} \|\varrho_\varepsilon D^\alpha u\| \\ &\leqslant C\Big(\|\varrho_\varepsilon(P(D)-z)u\| + \sum_{j=1}^{n} \|\varrho_\varepsilon P^{(j)}(D)u\|\Big) + \varepsilon K \sum_{|\alpha|\leqslant m-1} \|\varrho_\varepsilon D^\alpha u\|, \end{aligned}$$

where K_1 and K are constants not depending on ε or z. Finally, choosing $\varepsilon=\varepsilon_0=\min(1/2K, 1)$, it follows from (A.19) that

$$\sum_{|\alpha|\leqslant m'} \|\varrho_{\varepsilon_0} D^\alpha u\| \leqslant 2C\Big(\|\varrho_{\varepsilon_0}(P(D)-z)u\| + \sum_{j=1}^{n} \|\varrho_{\varepsilon_0} P^{(j)}(D)u\|\Big)$$

for all $u \in \mathcal{H}_{m,s}(\boldsymbol{R}^n)$ and $z \in \mathcal{K}$. This implies (A.13) and completes the proof of the lemma.

We pass now to the

Proof of Theorem A.1. Let $s>\frac{1}{2}$ and let $\mathcal{K}$ be a compact set in $\boldsymbol{C}\setminus\Lambda_C(P)$. By Lemma A.3 there exists a constant C_{-s} such that

$$\text{(A.20)} \qquad \|u_{m',-s}\| \leqslant C_{-s}\Big(\|(P(D)-z)u\|_{0,-s} + \sum_{j=1}^{n} \|P^{(j)}(D)u\|_{0,-s}\Big)$$

for all $u \in \mathcal{H}_m(\boldsymbol{R}^n)$ and $z \in \mathcal{K}$. By Lemma A.2, we have

$$\text{(A.20′)} \qquad \|P^{(j)}(D)u\|_{0,-s} \leqslant mc_s\|(P(D)-z)u\|_{0,s}, \qquad j=1,\dots,n$$

for all $u \in \mathcal{H}_m(\boldsymbol{R}^n)$ and $z \in \boldsymbol{C}$. Hence, combining (A.20) and (A.20′), we get

$$\|u\|_{m',-s} \leqslant C_{-s}\big(\|(P(D)-z)u\|_{0,-s} + nmc_s\|(P(D)-z)u\|_{0,s}\big) \leqslant C\|(P(D)-z)u\|_{0,s},$$

for all $u \in \mathcal{H}_m(\boldsymbol{R}^n)$ and $z \in \mathcal{K}$. This yields (A.2) and establishes the theorem.

Appendix B.

In connection with Theorem 3.2 we establish here the following more general result.

THEOREM B.1. *Let $F(x)$ be a real C^∞ function on $\boldsymbol{R}^n$ such that $F(x) \neq 0$ for $|x| \geqslant R_0$, and which satisfies*

$$\left| D^\alpha \left(\frac{1}{F(x)} \right) \right| \leqslant C_\alpha \qquad \text{for } |x| \geqslant R_0 \tag{B.1}$$

for all multi-indices α; R_0 and C_α certain constants. Set $\Gamma = \{x: F(x) = 0\}$. Assume that Γ is not empty and that grad $F(x) \neq 0$ *for $x \in \Gamma$, so that Γ is a smooth compact $n-1$ dimensional manifold in $\boldsymbol{R}^n$.*

Let now $u(x)$ be a function in $\mathcal{H}_s(\boldsymbol{R}^n)$, with $s > \frac{1}{2}$, such that $u(x) = 0$ on Γ (trace sense). Then,

$$\frac{u}{F} \in \mathcal{H}_{s-1}(\boldsymbol{R}^n) \cap L^1_{\text{loc}}(\boldsymbol{R}^n)\,, \tag{B.2}$$

and

$$\left\| \frac{u}{F} \right\|_{s-1} \leqslant C \|u\|_s \tag{B.2'}$$

where C is a constant depending only on F and s.

We note that Theorem 3.2 is an easy corollary of Theorem B.1, with $F(x) = |x|^2 - k^2$, using also the following well known result (e.g. [15], Ch. 1).

LEMMA B.1. *Let $a(x)$ be a bounded C^∞ function on $\boldsymbol{R}^n$ such that $\sup_x |D^\alpha a(x)| < \infty$ for $\forall \alpha$. Then the map: $u \to au$ takes $\mathcal{H}_s(\boldsymbol{R}^n)$ into itself for every real s. One has*

$$\|au\|_s \leqslant c_s \|u\|_s \qquad \text{for } u \in \mathcal{H}_s(\boldsymbol{R}^n)$$

where c_s is a constant depending only on a and s.

The main step in the proof of Theorem B.1 is given in the following

LEMMA B.2. *Let $u(x)$ be a function in $\mathcal{H}_s(\boldsymbol{R}^n)$ with $s > \frac{1}{2}$. Set $u(x) = u(x_1, x')$, $x' = (x_2, \ldots, x_n)$, and assume that $u(0, x') = 0$ (trace sense). Then,*

$$\frac{u}{x_1} \in \mathcal{H}_{s-1}(\boldsymbol{R}^n) \cap L^1_{\text{loc}}(\boldsymbol{R}^n) \tag{B.3}$$

and

$$\left\| \frac{u}{x_1} \right\|_{s-1} \leq \gamma_s \| u \|_s \tag{B.3$'$}$$

where γ_s is a constant depending only on s.

PROOF. Consider $u(x_1, \cdot)$ as a function of x_1 with values in $L^2(\boldsymbol{R}_{x'}^{n-1})$. It is well known (e.g. [15], Ch. 1) that $u(x_1, \cdot)$ is a Hölder continuous function satisfying

$$\| u(x_1, \cdot) - u(y_1, \cdot) \|_{L^2(\boldsymbol{R}^{n-1})} \leq c_s |x_1 - y_1|^{s-\frac{1}{2}} \| u \|_s ,$$

where c_s is a constant depending only on s. Since $u(0, \cdot) = 0$, we have

$$\| u(x_1, \cdot) \|_{L^2(\boldsymbol{R}^{n-1})} \leq c_s |x_1|^{s-\frac{1}{2}} \| u \|_s ,$$

which implies that

$$\frac{u(x)}{x_1} \in L^1_{\mathrm{loc}}(\boldsymbol{R}^n) .$$

Next we show that $u/x_1 \in \mathcal{H}_{s-1}(\boldsymbol{R}^n)$, assuming first that u is a function with compact support. Set

$$f(x) = \frac{u(x)}{x_1} \tag{B.4}$$

and note that by the preceding f is a function with compact support in $L^1(\boldsymbol{R}^n)$. Taking Fourier transform, it follows that $\hat{f}(\xi)$ and $\hat{u}(\xi)$ are C^∞ functions on $\boldsymbol{R}^n$, tending to zero as $|\xi| \to \infty$, and satisfying

$$\frac{\partial}{\partial \xi_1} \hat{f}(\xi) = -i \hat{u}(\xi) . \tag{B.4$'$}$$

We shall use the following inequality due to Hardy. Let $g(t)$ be a continuously differentiable function on $\boldsymbol{R}$ such that $g(t) \to 0$ as $t \to \pm \infty$. Then for any $s > \frac{1}{2}$,

$$\int_{-\infty}^{\infty} |g(t)|^2 t^{2s-2} \, dt \leq \left(\frac{2}{2s-1} \right)^2 \int_{-\infty}^{\infty} |g'(t)|^2 t^{2s} \, dt . \tag{B.5}$$

(One proves (B.5) by integration by parts, noting that when the right hand side of (B.5) is finite then $g(t) = o(|t|^{-2s+1})$ as $t \to \pm \infty$).

We shall write $\hat{f}(\xi)=\hat{f}(\xi_1,\xi')$, $\xi'=(\xi_2,\ldots,\xi_n)$, and shall apply (B.5) to the function $g(t)=\hat{f}(t,\xi')$. We get

$$\text{(B.6)}\qquad \int_{R^n}|\hat{f}(\xi)|^2\xi_1^{2s-2}\,d\xi\leqslant\left(\frac{2}{2s-1}\right)^2\int_{R^n}\left|\frac{\partial\hat{f}}{\partial\xi_1}\right|^2\xi_1^{2s}\,d\xi= =\left(\frac{2}{2s-1}\right)^2\int_{R^n}|\hat{u}(\xi)|^2\xi_1^{2s}\,d\xi\leqslant\left(\frac{2}{2s-1}\right)^2\|u\|_s^2<\infty\,.$$

If $\frac{1}{2}<s\leqslant1$, it follows from (B.6) and (B.4) that

$$\text{(B.6')}\qquad \left\|\frac{u}{x_1}\right\|_{s-1}\leqslant\frac{2}{2s-1}\,\|u\|_s\,,$$

which gives the desired result in the case considered. If $s>1$, an application of (B.5) to $g(t)=\hat{f}(t,\xi')$ with $s=1$ gives

$$\text{(B.7)}\qquad \int_{-\infty}^{\infty}|\hat{f}(\xi_1,\xi')|^2\,d\xi_1\leqslant4\int_{-\infty}^{\infty}|\hat{u}(\xi_1,\xi')|^2\xi_1^2\,d\xi_1\,.$$

Multiplying (B.7) by $(1+|\xi'|^2)^{s-1}$ and integrating with respect to ξ', we find

$$\text{(B.8)}\qquad \int_{R^n}|\hat{f}(\xi)|^2(1+|\xi'|^2)^{s-1}\,d\xi\leqslant4\int_{R^n}|\hat{u}(\xi)|^2\xi_1^2(1+|\xi'|^2)^{s-1}\,d\xi\leqslant4\|u\|_s^2\,.$$

Adding (B.6) and (B.8) it follows that

$$\text{(B.9)}\qquad \left\|\frac{u}{x_1}\right\|_{s-1}^2=\int_{R^n}|\hat{f}(\xi)|^2(1+|\xi|^2)^{s-1}\,d\xi\leqslant \leqslant\frac{\gamma_s^2}{8}\int_{R^n}|\hat{f}(\xi)|^2\left[\xi_1^{2s-2}+(1+|\xi'|^2)^{s-1}\right]d\xi\leqslant\gamma_s^2\|u\|_s^2$$

where γ_s is a constant depending only on s. This yields (B.3′) and establishes the lemma for u with compact support.

To prove (B.3′) in the general case, let $\zeta\in C_0^\infty(\boldsymbol{R}^n)$ be a cutoff function ($\zeta(x)=1$ for $|x|\leqslant1$) and set $u_j=\zeta(x/j)\,u$, $j=1,2,\ldots$. Then $u_j\in\mathcal{H}_s(\boldsymbol{R}^n)$ and by the result just proved $u_j/x_1\in\mathcal{H}_{s-1}\cap L^1_{\text{loc}}$ and the corresponding inequality (B.3′) holds. Letting $j\to\infty$ it follows readily (since $u_j\to u$ in $\mathcal{H}_s(\boldsymbol{R}^n)$ and $u_j/x_1\to u/x_1$ in $L^1_{\text{loc}}(\boldsymbol{R}^n)$) that $u/x_1\in\mathcal{H}_{s-1}(\boldsymbol{R}^n)$ and that (B.3′) holds. This yields the lemma.

A somewhat more general form of Lemma B.2 is the following

LEMMA B.2 BIS. *Let $g(x') \in C_0^\infty(\mathbf{R}_{x'}^{n-1})$, $x' = (x_2, \ldots, x_n)$. Set $\Gamma = \{x: x = (x_1, x') \in \mathbf{R}^n,\ x_1 - g(x') = 0\}$. Let $u(x)$ be a function in $\mathcal{H}_s(\mathbf{R}^n)$ with $s > \frac{1}{2}$ such that $u|_\Gamma = 0$. Then,*

$$\frac{u(x)}{x_1 - g(x')} \in \mathcal{H}_{s-1}(\mathbf{R}^n) \cap L^1_{\mathrm{loc}}(\mathbf{R}^n)$$

and

$$\left\| \frac{u}{x_1 - g(x')} \right\|_{s-1} \leqslant \gamma_s \|u\|_s ,$$

where γ_s is a constant depending only on s and g.

PROOF. By change of coordinates: $y_1 = x_1 - g(x')$, $y_i = x_i$ for $i = 2, \ldots, n$, one defines a C^∞ diffeomorphism $x \to y = \varphi(x)$ of $\mathbf{R}_x^n$ onto $\mathbf{R}_y^n$. By a well known result (e.g. [15], p. 94) the map $\varphi^*: u(x) \to u(\varphi^{-1}(y))$ is an isomorphism of $\mathcal{H}_s(\mathbf{R}^n)$ onto itself for every real s. The conclusion of the lemma follows now immediately from this result and from Lemma B.2 applied to the function $u(\varphi^{-1}(y))$ in $\mathbf{R}_y^n$. We give now the

PROOF OF THEOREM B.1. Using a partition of unity and Lemma B.1 it is readily seen that to prove the theorem it suffices to show that for every point $y \in \mathbf{R}^n$ there exists an open neighborhood U_y such that the conclusion of the theorem holds for all functions u satisfying the conditions of the theorem and having their support in U_y, and also that the same is true for a neighborhood of infinity $U_\infty = \{x: |x| > R_0\}$. (The constant C in (B.2)$'$ may depend on U_y.) To prove this result suppose first that y is any point in $\mathbf{R}^n$ such that $y \notin \Gamma$. In this case let U_y be any open neighborhood of y such that $\overline{U}_y \cap \Gamma = \emptyset$. Using a cutoff function we define a function $a(x) \in C_0^\infty(\mathbf{R}^n)$ such that $a(x) = 1/F(x)$ for $x \in U_y$. Since $u/F = au$ for any $u \in \mathcal{H}_s(\mathbf{R}^n)$ with $\operatorname{supp} u \subset U_y$, it follows from Lemma B.1 that $u/F \in \mathcal{H}_s(\mathbf{R}^n)$ and that

$$\left\| \frac{u}{F} \right\|_s \leqslant C \|u\|_s \tag{B.10}$$

where C is a constant independent of u. One sees similarly that (B.10) holds for all functions $u \in \mathcal{H}_s(\mathbf{R}^n)$ with $\operatorname{supp} u \subset \{x: |x| > R_0\}$, R_0 chosen sufficiently large.

Finally, let $y \in \Gamma$. Since $(\operatorname{grad} F)(y) \neq 0$, we may assume without loss of generality that $(\partial F/\partial x_1)(y) \neq 0$. By the implicit function theorem it follows that in a sufficiently small ball $B_\delta = \{x\colon |x-y| < \delta\}$, the function F admits a factorization

$$F(x) = G(x)\big(x_1 - h(x')\big)\,, \qquad x' = (x_2, \ldots, x_n)\,,$$

where $G \in C^\infty(B_\delta)$, $G \neq 0$ in B_δ, and where $h(x')$ is a C^∞ function of x' for $|x'-y'| < \delta$. Set $U_\delta = \{x\colon |x-y| < \delta/2\}$ and choose functions $a(x) \in C_0^\infty(\boldsymbol{R}^n)$ and $g(x') \in C_0^\infty(\boldsymbol{R}^{n-1})$ such that $a(x) = 1/G(x)$ in U_δ, and $g(x') = h(x')$ for $|x'-y'| < \delta/2$. For any $u \in \mathcal{H}_s(\boldsymbol{R}^n)$ with $\operatorname{supp} u \subset U_\delta$, we have

$$\text{(B.11)} \qquad \frac{u}{F(x)} = a(x)\,\frac{u}{x_1 - g(x')}\,.$$

If in addition $u|_\Gamma = 0$, it follows from (B.11) upon application of Lemma B.2 bis and Lemma B.1 that $u/F \in \mathcal{H}_{s-1}(\boldsymbol{R}^n) \cap L^1_{\text{loc}}(\boldsymbol{R}^n)$ and that (B.2)$'$ holds with a constant C depending only on a, g and s. This completes the proof of the theorem.

Appendix C.

We give here the proof of Lemma 5.1. Set $B_r = \{x\colon |x| < r,\ x \in \boldsymbol{R}^n\}$. It is readily seen that Lemma 5.1 is implied by the following local regularity result.

THEOREM C.1. *Let $u(x)$ be a function in $\mathcal{H}_2(B_1)$. Suppose that u verifies the differential equation*

$$\text{(C.1)} \qquad -\Delta u + q(x)\,u = f(x) \quad \text{in } B_1$$

where q and f belong to $L^2(B_1)$.

Suppose also that there exists a number θ, $0 < \theta < \frac{1}{2}$, and positive constants Q and F such that

$$\text{(C.2)} \qquad \sup_{x \in B_1} \int_{B_1} |q(y)|^2\,|y-x|^{-n-2\theta+4}\,dy \leqslant Q^2\,,$$

and

$$\text{(C.2}'\text{)} \qquad \sup_{x \in B_1} \int_{B_1} |f(y)|^2\,|y-x|^{-n-2\theta+4}\,dy \leqslant F^2\,.$$

Then u is a Hölder continuous function in B_1. For every $0 < r < 1$ the following estimate holds

$$\|u\|_{C^\theta(B_r)} \leqslant C_r(Q+1)^\nu(\|u\|_{L^2(B_1)} + F) \tag{C.3}$$

where ν is a positive constant depending only on θ and n, and C_r is a constant depending only on θ, n and r. Here

$$\|u\|_{C^\theta(B_r)} = \sup_{|x|<r} |u(x)| + \sup_{\substack{|x|<r\\|y|<r}} \frac{|u(x)-u(y)|}{|x-y|^\theta}\,.$$

We also have

$$\|u\|_{\mathcal{H}_2(B_r)} \leqslant C_r(Q+1)^\nu(\|u\|_{L^2(B_1)} + F)\,. \tag{C.3$'$}$$

PROOF. It suffices to prove (C.3) since (C.3′) follows from (C.3) and (C.1) by the interior L^2 estimates for solutions of the Laplace equation. For convenience we shall assume that $n \geqslant 3$. With an obvious modification our proof is also valid for $n = 2$.

Suppose first that the restriction of u to B_r $(0 < r \leqslant 1)$ is in $L^p(B_r)$ for some p, $2 \leqslant p < \infty$. Then we claim that $u \in L^{p_1}_{\text{loc}}(B_r)$ with $p_1 > p$, $p_1 \leqslant \infty$ given by

$$\frac{1}{p_1} = \max\left(\frac{1}{p} - \delta,\ 0\right) \tag{C.4}$$

where δ is a positive number depending only on θ and n (one may take $2\delta = \theta/(n-\theta)$). Moreover, for any $0 < r_1 < r$ we have the estimate

$$\|u\|_{L^{p_1}(B_{r_1})} \leqslant C(Q+1)(\|u\|_{L^2(B_r)} + F) \tag{C.5}$$

where C is a constant depending only on θ, n, r and r_1.

Assume for a moment the validity of the claim just made. Then, since $u \in L^2(B_1)$, it would follow from the above that $u \in L^{p_0}_{\text{loc}}(B_1)$ with $1/p_0 = (1/p - \delta)_+$, and that for any $0 < a < 1$

$$\|u\|_{L^{p_0}(B_a)} \leqslant C(Q+1)(\|u\|_{L^2(B_1)} + F)\,. \tag{C.6}$$

If $p_0 < \infty$, we apply again (C.5) with $p = p_0$, $r = a$ and $r_1 = a^2$, concluding that $u \in L^{p_1}(B_{a^2})$ with $1/p_1 = (1/p_0 - 1/\delta)_+ \leqslant (\frac{1}{2} - 2\delta)_+$, and that

$$\|u\|_{L^{p_1}(B_{a^2})} \leqslant C(Q+1)(\|u\|_{L^{p_0}(B_a)} + F)\,. \tag{C.6$'$}$$

Iterating this argument at most $\nu_0 = [1/2\delta] + 1$ times it follows that u is bounded in $B_{a^{\nu_0}}$. Combining the successive inequalities (C.6), (C.6′), ..., setting $a^{\nu_0} = r$ and noting that r may be any number < 1, we conclude that u is locally bounded in B_1 and that

$$\sup_{|x|<r} |u(x)| \leqslant \tilde{C}_r (Q+1)^{\nu_0} (\|u\|_{L^2(B_1)} + F) \tag{C.7}$$

where $\tilde{C}_r$ is a constant depending only on r.

Turning to the proof of (C.5), we denote by $E(x)$ the fundamental solution of $-\Delta$, $E(x) = \gamma_n |x|^{-(n-2)}$ with $\gamma_n = \frac{1}{4}\Gamma(n/2-1)\pi^{-n/2}$. We introduce the two functions:

$$w_0(x) = \int_{|y|<1} f(y)\, E(x-y)\, dy\,, \tag{C.8}$$

$$v_r(x) = -\int_{|y|<r} q(y)\, u(y)\, E(x-y)\, dy\,. \tag{C.8′}$$

Using (C.2′) it follows readily that $w_0(x)$ is a bounded function in B_1 satisfying

$$|w_0(x)| \leqslant cF \qquad \text{for } |x| < 1 \tag{C.9}$$

where c is a constant depending only on θ and n. Also, using (C.2), it follows that

$$|v_r(x)| \leqslant \gamma_n Q \Big(\int_{B_r} |u(y)|^2 |y-x|^{-n+2\theta} dy \Big)^{\frac{1}{2}} \tag{C.10}$$

for almost all x in B_1. Since $|u(y)|^2 \in L^{p/2}(B_r)$, and since $|x|^{-n+2\theta} \in L^{1/(1-2\delta)}(B_1)$, with $2\delta = \theta/(n-\theta)$, it follows from (C.10) applying Young's inequality that $v_r \in L^{p_1}(B_1)$ with $1/p_1 = (1/p - \delta)_+$ and that

$$\|v_r\|_{L^{p_1}(B_1)} \leqslant c_1 Q \|u\|_{L^p(B_r)} \tag{C.11}$$

where c_1 depends only on θ and n.

Set $h_r = u - v_r - w_0$. It follows from (C.1), (C.8) and (C.8′) that h_r is harmonic in B_r, so that in particular $h_r \in C^\infty(B_r)$. Using a standard estimate for harmonic functions, together with (C.11) and (C.9), we find that for any $r_1 < r$:

$$\begin{aligned} \|h_r\|_{C^0(B_{r_1})} &\leqslant c_2 \|h_r\|_{L^2(B_r)} \leqslant \\ &\leqslant c_2 \big(\|u\|_{L^2(B_r)} + \|v_r\|_{L^2(B_r)} + \|w_0\|_{L^2(B_r)} \big) \leqslant \\ &\leqslant c_3 \big((Q+1) \|u\|_{L^2(B_r)} + F \big)\,, \end{aligned} \tag{C.12}$$

where c_2, c_3 depend only on r, r_1, θ and n.

Finally, since

$$u = v_r + w_0 + h_r \,, \tag{C.13}$$

it follows from the above that $u \in L^{p_1}_{\mathrm{loc}}(B_r)$. More precisely, combining (C.13), (C.12), (C.11) and (C.9) we obtain (C.5), as claimed.

We have already shown that (C.5) implies that u is locally bounded in B_1. To show that u is Hölder continuous we use the following

LEMMA. *Let*

$$w(x) = \int_{|y|<1} g(y)\, E(x-y)\, dy$$

where g is a function in $L^2(B_1)$ such that

$$\sup_{|x|<1} \int_{|y|<1} |g(y)|^2\, |y-x|^{-n-2\theta+4}\, dy \leqslant G^2 < \infty \,,$$

$0<\theta<\frac{1}{2}$. *Then, $w \in C^\theta(B_1)$ and*

$$\|w\|_{C^\theta(B_1)} \leqslant c_0 G$$

where c_0 is a constant depending only on θ and n.

We omit the proof of the lemma which follows by a standard potential theoretic argument. To complete the proof of the theorem we use again the decomposition (C.13). Applying the above lemma we see that w_0 is Hölder continuous of order θ and

$$\|w_0\|_{C^\theta(B_1)} \leqslant c_0 F \,. \tag{C.14}$$

Also, by the same lemma applied to v_r it follows that $v_r \in C^\theta(B_1)$. More precisely, taking note of (C.8′), (C.7) and (C.2), we find that

$$\|v_r\|_{C^\theta(B_r)} \leqslant c_0 Q \sup_{|x|<r} |u(x)| \leqslant c_0 \tilde{C}_r (Q+1)^{r_0+1} \big(\|u\|_{L^2(B_1)} + F \big) \,. \tag{C.14′}$$

Combining (C.13), (C.12), (C.14) and (C.14′), it follows that $u \in C^\theta(B_r)$ and that (C.3) holds. This establishes the theorem.

REFERENCES

[1] S. Agmon, *Lectures on elliptic boundary value problems*, Van Nostrand, Math. Studies, **2** (1965).

[2] P. Alsholm - G. Schmidt, *Spectral and scattering theory for Schrödinger operators*, Arch. Rational Mech. Anal., **40** (1971), pp. 281-311.

[3] N. Dunford - J. T. Schwartz, *Linear operators*, Part II, Interscience, New York - London, 1963.

[4] U. Greifenegger - K. Jörgens - J. Weidmann - M. Winkler, *Streutheorie für Schrödinger-Operatoren*, to appear.

[5] T. Ikebe, *Eigenfunction expansions associated with the Schrödinger operators and their applications to scattering theory*, Arch. Rational Mech. Anal., **5** (1960), pp. 1-34.

[6] T. Kato, *Growth properties of solutions of the reduced wave equation with a variable coefficient*, Comm. Pure Appl. Math., **12** (1959), pp. 403-425.

[7] T. Kato, *Perturbation theory for linear operators*, Springer, 1966.

[8] T. Kato, *Some results on potential scattering*, Proc. International Conf. Functional Analysis and Related Topics, Tokyo, 1969, Tokyo University Press, 1970, pp. 206-215.

[9] T. Kato, *Scattering theory and perturbation of continuous spectra*, Proc. International Congress Math., Nice, 1970, Gauthier-Villars, vol. 1 (1971), pp. 135-140.

[10] T. Kato, S. T. Kuroda - *Theory of simple scattering and eigenfunction expansions*, Functional Analysis and Related Fields, edited by F. E. Browder, Springer, 1970, pp. 99-131.

[11] S. T. Kuroda, *On the existence and the unitary property of the scattering operator*, Nuovo Cimento, **12** (1959), pp. 431-454.

[12] S. T. Kuroda, *Spectral representations and the scattering theory for Schrödinger operators*, Proc. International Congress Math., Nice, 1970, Gauthier-Villars, vol. 2 (1971), pp. 441-445.

[13] S. T. Kuroda, *Scattering theory for differential operators - I: Operator theory*, J. Math. Soc. Japan, **25** (1973), pp. 75-104.

[14] S. T. Kuroda, *Scattering theory for differential operators - II: Self-adjoint elliptic operators*, J. Math. Soc. Japan, **25** (1973), pp. 222-234.

[15] J. L. Lions - E. Magenes, *Problèmes aux limites non homogènes et applications*, vol. 1, Ed. Dunod, Paris, 1968.

[16] J. Milnor, *Singular points of complex hypersurfaces*, Ann. Math. Studies, vol. 61, Princeton University Press and the University of Tokyo Press, Princeton, 1968.

[17] A. Ja. Povzner, *The expansion of arbitrary functions in terms of eigenfunctions of the operator* $-\Delta u + cu$, Math. Sbornik, **32** (1953), pp. 109-156; A.M.S. Translations, Series 2, **60** (1967), pp. 1-49.

[18] P. A. Rejto, *Some potential perturbations of the Laplacian*, Helvetica Physica Acta, **44** (1971), pp. 708-736.

[19] M. Schechter, *Spectra of partial differential operators*, North Holland, 1971.

[20] M. Schechter, *Scattering theory for elliptic operators of arbitrary order*, Comm. Mat. Helvetici, **49** (1974), pp. 84-113.

[21] J. R. SCHULENBERGER - C. H. WILCOX, *Eigenfunction expansions and scattering theory for wave propagation problems of classical physics*, Arch. Rational Mech. Anal., **46** (1972), pp. 280-320.
[22] N. SHENK - D. THOE, *Eigenfunction expansions and scattering theory for perturbations of* $-\Delta$, The Rocky Mountain J. of Math., **1** (1971), pp. 89-125.
[23] D. THOE, *Eigenfunction expansions associated with Schrödinger operators in* R^n, $n \geqslant 4$, Arch. Rational Mech. Anal., **26** (1967), pp. 335-356.